ENERGY

ENE

RGY

The New Agenda

RAYMOND M. WRIGHT

First published in Jamaica, 2010 by
Raymond M. Wright
1 Great House Mews
Kingston 6
Jamaica, West Indies

Raymond.Wright@pcj.com

National Library of Jamaica Cataloguing in Publication Data

Wright, Raymond M.
  Energy : the new agenda / Raymond M. Wright

     p. : ill. ; cm.

Bibliography : p.   .– Includes index.

ISBN: 978-976-8217-82-0

1. Power resources – Jamaica 2. Renewable energy sources
3. Energy development
I.   Title

333.79   – dc 22

Set in Sabon 10.5/15 x 24.

Cover and book design by Robert Harris.
(E-mail: roberth@cwjamaica.com)

Printed and bound in the United States of America.

TO THOSE COMMITTED TO DEVELOPING

A GREENER ENERGY ECONOMY

THROUGH A NEW ENERGY AGENDA

.

Contents

List of Tables

List of Figures

Foreword

ENERGY IS LIFE, FOOD AND WATER. Energy is progress and riches. Energy is politics and security.

Energy is the foundation stone of all modern economies and it is instructive to see how the discovery and exploitation of all energy resources have shaped our current world. It is vitally important to fully understand the history of energy in order to equip ourselves to plan for and achieve future prosperity without destroying along the way the environment on which all societies depend.

In 'Energy in the service of society: prospects and challenges' we are presented with Raymond Wright's major thesis that energy security is fundamental to the economic growth, health and prosperity of every nation. He asserts that a nation must diversify its energy sources and reduce its dependence on conventional fossil fuels through the expansion of renewable energies. He also looks to the future of such disparate energies as nuclear fission, fuel cells, Micro and Pico-hydro to waste-to-energy and the more conventional renewables. While not ignoring issues such as population growth, reduction of arable land for agriculture, and mass migration to cities and immigration; he identifies the major policies and incentives which will be required if we are to achieve a cleaner and sustainable environment. It is predicted that by 2050 two-thirds of the world population will be living in urban settlements. Already, modes of transportation and excessive use of coal for electricity generation has caused irretrievable damage to the environment and health. To add to this complex situation we now have the fact of global recession which could be partially stabilised through the creation of millions of jobs in renewable energy.

'The power of oil: geopolitics, economic gain and addiction to the world's primary fuel'; as would be expected of an oil man, Raymond

Wright has given us a very interesting account of the rise to power of the oil companies and oil rich countries. He writes knowledgeably on the rise of OPEC as well as the more recent shift towards production from the non-OPEC sources. A case study of the oil exploration history of Jamaica is given as an example of exploration activity in a frontier province. Similar in-depth analyses are provided for both the coal and the natural gas industries. It is perhaps more interesting to consider the nuclear energy industry because here there is more of a divide over the benefits and dangers associated with this energy source. The two forms of nuclear energy, fission and fusion are discussed. At the time of writing there are 450 active reactors in operation in 30 countries which contribute 15 per cent to global electricity supply, full details are helpfully tabulated. Importantly, risk, and its perception, is addressed based on the number and nature of accidents and the vexed problem of waste disposal. Clearly, this is an area which all governments need to discuss more fully before committing themselves to the huge investment that nuclear energy entails. It must not be forgotten that compared to all other forms of energy, nuclear energy is a relatively costly source when factoring in risks and waste disposal. However, the nuclear energy industry is not static and there are many possible developments in the fission and small-scale plants which may be the way forward in the long run.

Raymond Wright devotes several chapters to discussion of the major renewable energy sources. The obvious success stories of photovoltaics, wind energy, biomass and hydropower are assessed and their future role and development are seen as viable sources of energy to replace fossil fuels. The new and exciting waste-to-energy and algal biomass are explained and evaluated and the success story of Brazil is used to illustrate how rapidly new sources can become important.

It is satisfying to see that the potentials of ocean and geothermal energy are recognised as they are so often overlooked when renewable energies are being discussed. They may well provide substantial future energy supplies. The complex field of hydrogen energy and fuels cells is elucidated and future scientific and cost breakthroughs are optimistically predicted.

Of course none of these energy scenarios make any sense unless and until we learn to reduce our energy demand and adopt energy conservation and efficiency through energy demand management and the production of efficient equipment and appliances. The author points out that energy conservation is often regarded as being equivalent to an energy source because it extends the life of conventional (and non-conventional) resources. The pivotal roles of the energy planner and architect are crucial for the design of sustainable buildings and communities. Topics such as passive buildings, daylighting and natural cooling are examined. The author believes that the long lead time required for all the fuel source transitions implied in this work has important implications for global energy strategies.

Finally, Raymond Wright presents a Renewable Energy Manifesto emphasizing the need for National Energy Policies, which will integrate both energy supply and energy efficient options into an overall strategy to meet energy service needs. He reminds us that governmental policies and public opinion are slow to change, and that melting ice-caps and glaciers, lowland flooding and changing weather patterns will lead to catastrophe for the earth's inhabitants unless we urge change NOW. Sustainable energy policy MUST prevail.

This book is a masterly compendium of knowledge and guidance for all those interested in energy policy and all those who care about the environment and our sustainable future.

PROFESSOR ALI SAYIGH
Chairman of World Renewable Energy Congress (WREC) and
Director General of World Renewable Energy Network (WREN)

April 2010

Acknowledgements

I WOULD LIKE TO ACKNOWLEDGE my editor Dr Angela Ramsay who carefully and critically edited the manuscript and in the process had me rewrite some concepts and many sentences.

This book would not have been possible without the zest, enthusiasm and competence of Gina-Lee Lawrence who guided the manuscript from initial ideas to final construct. She should be applauded for her dedication to the task.

Mr Robert Harris showed his strong artistic talents in the design and production of the book. Kim Hoo Fatt exhibited proficient proof reading services.

To these persons, and to others unnamed, I express my sincere gratitude.

It is hoped that the book will contribute to accelerating the move to greater utilization of cleaner energy resources. In spite of the help I have received in its preparation and production, I am solely responsible for any errors or faults which may be present.

Abbreviations

ae	alternating current
AFC	Alkaline Fuel Cells
b/d	barrels of oil per day
BOD	biochemical oxygen demand
Btu	British thermal unit (1 Btu = 1055.06 joules)
CCS	carbon capture and sequestration
CdT	ecadmium telluride
CERA	Cambridge Energy Research Associates
CIS	copper indium diselenide
CNG	Compressed Natural Gas
CO_2	carbon dioxide
dc	direct current
DOE	Department of Energy (USA)
DMFC	Direct Methanol Fuel Cells
DSM	Demand-Side Management
EC	European Community
GaAs	gallium arsenide
GEF	Global Environment Facility
GHG	greenhouse gases
Gtoe	gigatonnes oil equivalent (109 tonnes oil equivalent)
GWh	gigawatt-hour (1,000,000 kilowatt-hours)
ha	hectare (= 0.01 square kilometres)
IEA	International Energy Agency
IGHAT	Integrated Gasification Humid Air Turbine

kV	kilovolts
kWh	kilowatt-hours
kWp	peak kilowatt
LNG	Liquefied Natural Gas
LPG	liquefied petroleum gas
M	metre
MCFC	Molten Carbonate Fuel Cells
Mtoe	megatonnes of oil equivalent (106 tonnes oil equivalent)
MTBE	Methyl Butyl Ether
MW	megawatt
MWp	peak megawatt
MSW	municipal solid waste
NOx	oxides of nitrogen
OTEC	Ocean Thermal Energy Conversion
O&M	Operating and Maintenance
OPEC	Organization of Petroleum Exporting Countries
OUR	Office of Utility Regulation
PAFC	Phosphoric Acid Fuel Cells
PCJ	Petroleum Corporation of Jamaica
PEM	Polymer Electrolyte Membrane
PV	photovoltaics
PWEC	Pelamis Wave Energy Converters
R&D	Research and Development
R3M	Remote Reservoir Resistivity Mapping
Si	silicon
SOFC	Solid Oxide Fuel Cells
TOE	tonnes of oil equivalent
TW	terawatt
UWI	University of the West Indies
V	volts
WEC	World Energy Council
WTE	Waste-to-Energy

Prologue

The production and use of energy causes more environmental damage than any other single economic activity.

– The *Economist*

The most critical technologies for sustainable development are energy technologies . . . The generation and use of energy are responsible for a large portion of almost all forms of pollution. For this reason alone, sustainable development will be impossible without new energy technologies.

– World Wildlife Fund

THE PURPOSE OF THIS BOOK is to educate readers about new energy generating possibilities. This book also provides an overview of conventional energy sources, namely oil, gas, coal and nuclear, in order to provide a context for the promotion of renewables. The other resources covered in the text include wind, solar photovoltaic, solar thermal, bioenergy, hydroelectricity, ocean energy, geothermal, and the hydrogen economy. Sustainable architecture, energy efficiency and future energy supply options are also discussed.

The energy sector presents problematic areas that conventional sources of supply have difficulties in complying with or addressing positively. For example, there is increasing concern about greenhouse gases, global warming and acid rain, making carbon based fuels such as oil and coal less acceptable for long-term use. For this reason and others, regions and countries such as those of the European Union, the USA and China are placing important emphasis on the utilization of renewables. Oil and coal supplies are gradually being depleted, emphasizing the long-term need for energy alternatives even if the environmental issues are resolved.

Resource	Years Remaining with Production at Present Rate (an approximation)
Crude oil	50
Natural Gas	80
Tar Sands	150
Coal	150
Uranium	50
Hydropower	Extensive, virtually unlimited (renewable)
Geothermal	Extensive, virtually unlimited (largely renewable)
Wind	Extensive, virtually unlimited (renewable)
Solar power	Extensive, virtually unlimited (renewable)
Bioenergy	Extensive, virtually unlimited (renewable)

Energy continues to be as important to our lives as air, water and earth. But the world is not about to run out of energy because there are large known reserves of conventional and renewable energy, as shown above.

As much as 25 per cent of the world's electricity production could come from renewable sources by 2030. Notably, over the past 15 years there has been a significant rise in the production of biofuels. On the other hand, the predicted growth of fuel cell technology has not been as rapid as initially perceived due to factors such as cost and size of engines for the transportation sector. A hydrogen fuel economy will probably come into being by about 2030. At that time small nuclear reactors will merge into the global energy mix in small scale plants producing 70–200 MW of energy. By that time, we would have begun to see the decline in the use of oil, both as a transport fuel and as a fuel for electricity generation. Oil will become a marquee fuel used primarily for petro chemicals.

Although renewables reduce the problems of global environmental change and resource depletion, they introduce new and special problems of their own. In some cases, their use creates local environmental problems, but even more difficult may be the changes in the culture and technology that may be necessary. For example, wind energy is intermittent, giving risk-averse utility companies some unfamiliar problems to face. With generation being dictated by factors such as changing weather, the electricity is not always produced and dispatched when needed. There will be increased reserved costs to compensate for periods of low power generation, or for the cost of increased storage: however, interconnection on the grid can help to smooth out the variability of intermittent power generation. The book provides many other examples of how problems associated with renewable energy may be resolved.

To effect a gradual transition to renewables, there is a need to enhance knowledge of renewable technologies both within the utility company, the regulatory body and civil society. For example, renewable sources of energy tend to be distributed in various locations and thus are a part of a general trend towards decentralized generation of power. With the increase in generation from renewables, utility companies may find themselves being not only generators and distributors but also coordinators of independent dispersed generation. The companies need to sensitize consumers about the association between renewable energy and environmental health. Indeed, reversing environmental degradation through education is an underlying theme throughout the entire book.

Knowledge about the environment is critical. Man uses the environment almost as a disposable article. He interferes with the environment yet understands little about its natural, complex design code, which he sometimes operates blindly. Accepting the limits of environmental space, there is a need for a new approach in the ecology debate. Until now, environmental policy has focused primarily on the disposal and prevention of wastes and pollutants. The quantities of energy used and substances moved, as well as increased land use create a problem in itself. An eco-efficiency revolution has to be induced, with resource saving integrated into the value chain, and education of populaces as a critical factor. The public needs to know that renewable energy helps their economies. There is now a growing consensus that renewable energy is central to reducing poverty and hunger for the poor, fostering social cohesion and economic growth. There is usually far less waste of resources when renewable energy is effectively used. For example, the increased use of renewables tends to come in conjunction with aggressive conservation and efficiency measures.

This book is useful for energy specialists, policy makers, educators and students in developed and developing countries. Jamaica is highlighted in the book as an example of a developing nation which has to import fossil fuels to satisfy most of its energy needs. In Jamaica, it is possible to increase the annual production of electricity from renewable sources to 20 per cent by the year 2020. The book provides suggestions on how Jamaica will be able to adapt and adopt those renewable energy technologies that are economic in the context of its energy mix.

Fortunately for Jamaica and other countries, the renewable energy industry is growing, looking to markets more than to governments. Regulators in the energy sector have become important in making clear rules and identifying market abuses. Once there is clarity about governmental goals and decision making, domestic and foreign private investment will follow. The global energy market is strong, growing, attracting private capital and shows positive indicators for the future.

1. | Energy in the Service of Society

Prospects and Challenges

To reduce our dependency on foreign oil, we need to develop all economic alternative energy resources.

– Barack Obama, 2009

ENERGY SECURITY IS FUNDAMENTAL TO the economic health and future of any country.

The French writer Antoine de Saint-Exupéry stated that "As for the future, your task is not to foresee it, but to enable it." Although we cannot accurately predict the future, there are at least five fundamental certainties. First, there will be a growth in demand for energy, primarily in sections of Asia. Second, there will be a need to diversify energy sources away from conventional fossil fuels with an expansion of cleaner energy from renewables. Third, the nuclear energy option will be reinvigorated because it will be cost effective and have little adverse effect on climate change. Fourth, over the short and medium term there will be an expansion in the use of coal accompanied by an increase in the production of greenhouse gases (GHG). Lastly, energy deficient countries would be faced with the economic burden of imported energy for decades while they develop to the maximum economic capacity domestic resources such as wind, solar, geothermal and hydropower.

The security of supply and volatility in petroleum prices have become important economic and political issues since 2000. Long-term measures to increase energy security centre on diversification, reducing dependence on any one source of imported energy, increasing the number of suppliers,

Electricity has enhanced the quality of life globally

exploiting native energy resources and reducing overall demand through energy conservation measures. Measures can also involve entering into international and regional agreements to strengthen energy trading relationships. Examples of agreements are the Energy Charter Treaty in Europe and the PetroCaribe initiative in the Caribbean.[1]

Renewable energy does not only enable greater energy security for developing countries; in most cases it is also coincident with clean energy. Presently, fossil fuels are the primary global energy source and also the primary source of pollution. Greenhouse gas emissions have stimulated climate change, which is one of the challenges facing the world today. Renewables, as part of a cleaner energy mix, are therefore rising in importance. Renewables cover a wide spectrum of energy sources. They include but are not limited to wind, solar, hydropower, geothermal, biomass, biofuels, and ocean energy. The use of renewables in the energy mix will be driven in part by public policy as well as the participation of the private sector in developing renewable energy sources as a business. A partnership between government and civil society is also required to help ensure that energy resources are used wisely.

In the context of energy deficient countries, renewable energy technologies will reduce fuel import cost as well as emissions. The goal is to increase the percentage of renewables in each country on an annual basis towards a reachable target. In some countries, a target in the range of 15–25 per cent of energy generated can be accommodated by renewables.

Important steps in a renewable energy strategy are:

- the creation of a national energy policy emphasizing the use of all economically viable renewable resources
- the encouragement of the use of renewable energy through tax incentives
- the promotion of energy efficiency and conservation
- the setting of reachable targets, obligatory or non-obligatory, for the development of renewable energy in each nation's energy mix
- the utilization of carbon trading as a vehicle to encourage the development of renewables
- the promotion of conservation and efficiency measures to slow or reduce high levels of energy intensity

Energy intensity is a measure of the energy efficiency of a nation's economy. It is calculated as units of energy per unit of gross domestic product (GDP). High energy intensity indicates that a country needs more energy consumption to generate one dollar of GDP, and low energy intensity indicates that the country needs less energy consumption to generate one dollar of GDP. Energy intensity is high in many developing nations, and the Caribbean serves as an interesting example. Table 1.1 shows the relative energy intensity of selected Caribbean countries.

Table 1.1   Energy Intensity of Selected Countries

Country	Energy Intensity
Guyana	7.2
Trinidad and Tobago	4.8
Haiti	4
Suriname	3.5
Jamaica	3
Dominican Republic	1
Barbados	0.9

Table 1.2   Relative Pricing Structure of Selected Caribbean Countries

Country	US$/kWh
Jamaica	0.27
Guyana	0.23
Barbados	0.22
Dominican Republic	0.21
Surinam	0.19
Haiti	0.09
Trinidad and Tobago	0.044

Table 1.2 shows the relative pricing structure of selected Caribbean countries. Jamaica has the highest electricity prices at 0.27 US cents per kWh, and Trinidad and Tobago has the lowest at 0.044 US cents per kWh.

The information on Tables 1.1 and 1.2 has been obtained primarily from an International Development Bank (IDB) survey released on September 15, 2008. The study concludes that energy efficiency and conservation are the most effective strategies through which Caribbean countries can conduct energy management with the greatest cost/benefit ratios. The study convincingly argues that capital spent on energy efficiency and conservation is significantly less than the capital required for the expansion of electricity generation facilities with a ratio of approximately 1:3. Thus energy efficiency is the most cost effective way of bolstering energy supply and energy security for an increasing population.

The world population is now 6.8 billion and by 2050 it will be approximately 9 billion, according to the United Nations Population Division. The amount of arable land is decreasing while the population is increasing. By 2030, more people in the developing world will live in urban than rural areas and by 2050, two-thirds of the global population is likely to be urban. Because people living in urban centres consume more energy than those in rural villages, there will be an attendant increase in energy consumption globally on a per capita basis. This new urban–rural mix will exacerbate the conflict and competition between food supply and fuel supply. There will be a need to harmonize pantry politics (the politics of food) with petro politics (the politics of fuel). It is incumbent on all countries to put in place long-term plans for their

energy future and the security of fuel supplies that ensure long-term supply, affordability and minimal impact on the environment.

There are sufficient reserves of various energy resources to last at least several decades at present rates of use if technologies with high energy-conversion efficient designs are utilized. The challenge is to use these resources in an environmentally acceptable manner while providing for the needs of growing populations and developing economies.[2] Conventional oil reserves will eventually peak as will natural gas reserves, but it is uncertain exactly when and what will be the nature of the transition to alternative liquid fuels such as coal-to-liquids, gas-to-liquids, oil shales, tar sands, heavy oils, and biofuels. It is still uncertain how and to what extent these alternatives will reach the market and what the resultant changes in global GHG emissions will be.

Conventional natural gas reserves are more abundant in energy terms than conventional oil, but in a commercial sense, are distributed less evenly across regions. Unconventional gas resources are also abundant, but the future economic development of these resources is uncertain. Coal, although abundant, is unequally distributed. Coal can be converted to liquids, gases, heat and power, although more intense utilization will demand viable carbon capture and sequestration (CCS) technologies if GHG emissions from its use are to be reduced. Nuclear energy, currently at approximately 7 per cent of total primary energy, could make an increasing contribution to carbon free electricity and heat in the future.

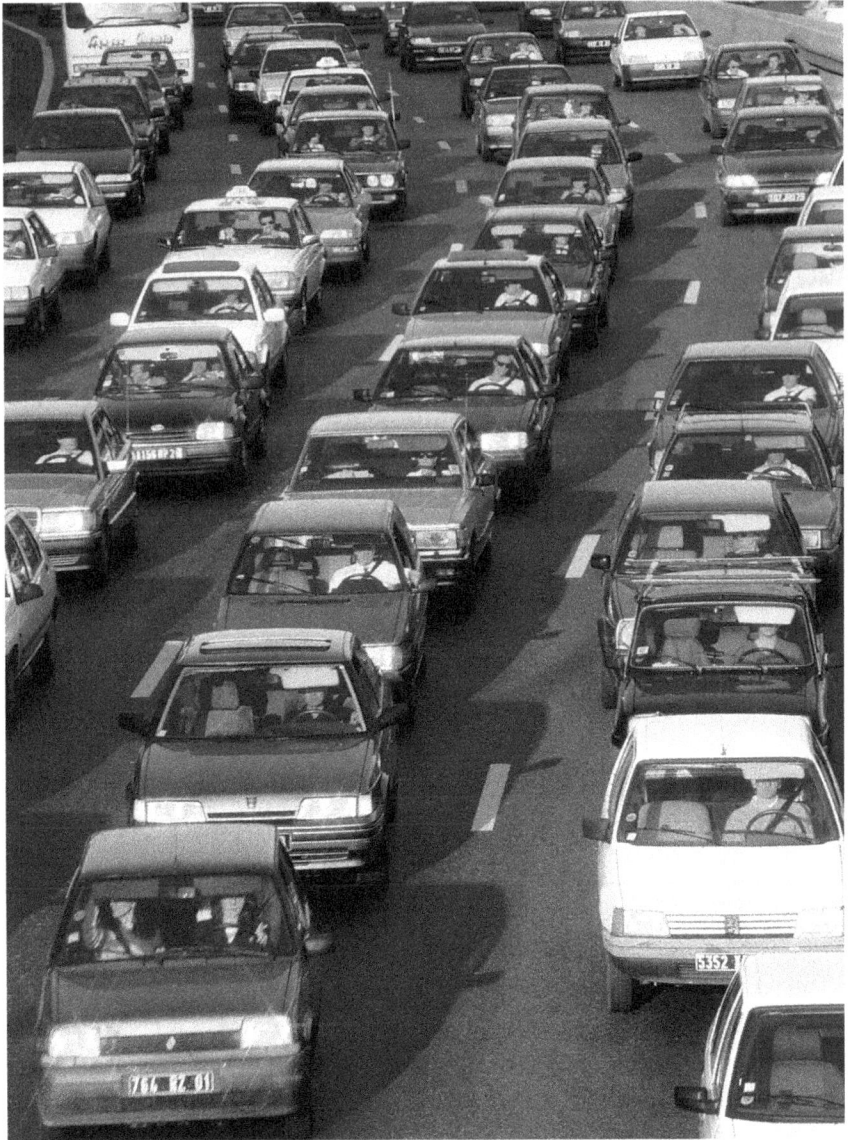

Among the major barriers to the expansion of nuclear energy are long-term fuel resource constraints without recycling, economics, safety, waste management, security and proliferation.

Renewable energy sources, with the exception of large hydro, are widely dispersed relative to fossil fuels, which are concentrated at individual locations that require major transport and distribution activities. Renewable energy provided in a distributed manner can result in savings in distribution and transmission costs. Most renewable energy-supply technologies are presently small overall contributors to global heat and electricity supply. Costs, as well as social and environmental concerns, are hurdles to the growth of renewable technology. For this reason, increased rates of deployment may need supportive government policies and strategies: for example, for the promotion of modern biomass. Traditional biomass for domestic heating and cooking now accounts for nearly 10 per cent of global energy supplies but could eventually be replaced; primarily by modern biomass and other renewable energy systems as well as by conventional fossil-based domestic fuels such as liquefied petroleum gas (LPG) and to a lesser extent kerosene.

The security of energy supply requirements and perceived financial benefits from strategic investments may not necessarily encourage the greater uptake of lower carbon-emitting technologies. Concerns about the future security of conventional oil, gas and electricity supplies could aid the transition to more low-carbon technologies such as nuclear, renewables and CCS. Furthermore, these same concerns could also encourage the greater uptake of unconventional oil and gaseous fuels as well as increase demand for coal and lignite in countries with abundant national supplies which are seeking national energy-supply security. Mitigating environmental impacts are usually predicated on the introduction of regulations and tax incentives rather than relying on market mechanisms. Small-scale, distributed energy plants using local energy resources and lower zero-carbon emitting technologies can give added reliability, be built more quickly and be efficient by utilizing both heat and power outputs locally. Distributed electricity systems can help reduce transmission losses and offset the high investment costs of upgrading distribution networks that are mature to full capacity. The provision of efficient energy services for all citizens in an environmentally sound way will require major investments in the energy-supply chain, conversion technologies and infrastructure (Sims et al., 2007).

In the final analysis, the world is now on track to achieve a sustainable energy future. The global energy supply will continue to be dominated by fossil fuels for several decades. The reduction of the resultant GHG emissions will require a transition to low-carbon technologies. This can happen over time as business opportunities and co-benefits are identified and responded to positively. The accelerated deployment of low-carbon fuel technologies will require policy intervention with respect to the complex and interrelated issues of security of energy supply, removal of structural advantages for fossil fuels, reducing related environmental impacts and achieving the challenging goals for sustainable development.

It might take some time for the world to recover from the global recession which began in 2008. As a number of economic stimulus packages are introduced to counter the effects of the global recession, the time is opportune to spend funds that generate growth, not only on infrastructure development but also on the provision of a greener low carbon energy future. This will not only have economic benefits but will also help to mitigate climate change. The money being spent today will have to be paid for by our children and perhaps even our children's children. We should not allow the next generation to blame us for not making the right choices.

References

Gilder, G. (1993) *Wealth and Poverty.* San Francisco: ICS Press.

International Development Bank (IDB) "How to Save US$36 Billion Worth of Electricity". A Survey of Energy Productivity in the Americas, released September 15, 2008.

Population Reference Bureau "Urban Population and Health in Developing Countries". Retrieved from: http://www.prb.org/Publications/Population Bulletins/2009/urbanization.aspx

Sims, R.E.H., Schock, R.N., Adegbululgbe, A., Fenhann, J., Konstantinaviciute, I., Moomaw, W., Nimir, H.B., Schlamadinger, B., Torres-Martínez, J., Turner, C., Uchiyama, Y., Vuori, S.J.V., Wamukonya, N., Zhang, X. (Eds.) "Energy Supply". In *Climate Change 2007: Mitigation.* Contribution of Working Group III to the Fourth Assessment Report of the Intergovernmental Panel on Climate Change. Cambridge: Cambridge University Press, Chapter 4.

United Nations Population Division "World Population Prospects: the 2008 Revision Population Database". Retrieved from: http://esa.un.org/unpp/index.asp.

2. | The Power of Oil

*Geopolitics, Economic Gain, and
Addiction to the World's Primary Fuel*

THE USE OF FOSSIL FUELS to satisfy almost all of the world's energy needs is a result of the long and fascinating history of oil. Oil seepages were tapped in Mesopotamia, Egypt and Persia as far back as 3000 BC and used for heating, roadmaking and building. The most famous source was Hit, on the river Euphrates near Babylon. Bitumen was a traded commodity in the ancient Middle East, used as building mortar to bind the walls of Babylon and Jericho. The medicinal and pharmaceutical value of oil was known to the Romans, and the Roman naturalist Pliny in the first century AD stated that oil checked bleeding, healed wounds, cured toothache, treated cataracts, stopped diarrhoea, and relieved rheumatism and fever. These are similar symptoms for which oil was used in the USA during 1850–1880.

Though the use of petroleum has a long history, the knowledge of its application was lost to the western hemisphere for many centuries, possibly because the known major sources of bitumen lay eastward beyond the boundaries of the Roman Empire. One of the earliest reports of a bitumen source in the Americas was made by Sir Walter Raleigh who visited the Trinidad pitch lake in 1595.[1]

Oil was first exploited for commercial use in the early 1850s when a group of men led by the entrepreneurial vision of George Bissell brought to market the inexpensive high-quality lamp illuminant that was so desperately needed in the mid-nineteenth century. The illuminant was derived from a process developed in 1850 by James Young of Glasgow,

On January 10, 1901, in Beaumont, Texas, the discovery of the first real oil gusher, called Spindletop, launched the oil boom

Scotland. 'Rock oil', as it was then termed, was also used to lubricate the moving parts of the dawning mechanical age. A cast of lively characters in Britain and North America carried the matter forward, defining a market and developing the refining technology upon which the oil industry would later be based. Dr Abraham Gesner developed a process for extracting oil from asphalt or similar substances and refined it into a high quality illuminating oil, which he called kerosene. Gesner helped to establish kerosene works in New York City and Boston. Parallel refining industries were developed in Britain based on low grade coal, and in France based on shale rock. By 1859 an estimated thirty-four companies in the United States were producing kerosene or 'coal-oil' as the product was generically known. The growth of this 'coal-oil' business, wrote the editor of a trade journal, was proof of "the impetuous energy with which the American mind takes up any branch of industry that promises to pay well" (Yergin, 1991).

In the 1850s, the use of kerosene in America faced two barriers. A substantial source of supply had yet to be found, nor was there a cheap lamp well-suited to burning the available kerosene. When that cheap lamp was created soon later in Vienna, it was exported to the United States. This kerosene lamp had a glass chimney and did not emit either smoke or an odour, and became the basis of the kerosene lamp trade in the United States and was later re-exported around the world.[2]

By 1859, oil was used from hand dug wells and other brine productions but 'Colonel' Edwin Drake arguably drilled the first commercial well that actually targeted oil. With the availability of an inexpensive lamp that could satisfactorily burn kerosene, Bissel and his fellow shareholders in the Pennsylvania Rock Oil Company tried to discover a new source of the raw material for which there was an established refining process. They wanted to find 'rock oil' cheaply and in abundance. Digging for oil was a slow process, and they felt that drilling would be a more efficient extraction method. The technique they used, known as salt-boring, had first been developed over fifteen hundred years earlier in China, but Bissell and his fellow investors adapted the salt-boring technique directly to oil. They hired Drake to arrange the drilling of the first well in Pennsylvania. Drake found a blacksmith named William Smith who undertook the drilling. On Saturday, August 27, 1859, at 69 feet, the drill dropped into a crevice underground and oil was found in Titusville.

With the resurgence of drilling as an extraction method, there was no shortage of rock oil. The only shortage was the whisky barrels to store

The Kerosene Lamp revolutionalized lighting and had worldwide applications

The discovery of commercial oil started a new industry and employment of new skills

the oil, and soon the barrels costed twice as much as the oil contained within them. Drake's success in drilling the first oil well would ultimately bequeath mobility and power to the people of the world. The drilling of that first well would also play a central role in the rise and fall of nations and empires and become a major element in the transformation of human society. There will perhaps never be a consensus regarding the claim that Drake drilled the first oil well, due to rival candidate wells from other geographic regions, but Drake stands alone in terms of the economic significance of his actions and the level of historical documentation and publicity which followed its efforts.

Other important developments occurred during this era. In 1865 in Cleveland, Ohio, two senior partners in the city's largest oil refinery resolved a dispute over the pace of expansion by holding a private auction between them, with the agreement that the higher bidder would own the company. John D. Rockefeller bid $72,500 and won. Fifty years after his successful bid, he said, "I ever point to that day as the beginning of the success I have made in my life." (Aeseng, 2000: 27). Thus the Standard Oil Company was born in 1870,[3] destined to become the largest oil producer in the world, dominating not only the oil business in the United States but also in the world.

At the end of the nineteenth century, the demand for artificial light was met by kerosene, gas and candles. In 1879, Thomas Edison developed the heat-resistant, incandescent light bulb and demonstrated lighting by electricity in 1882.[4] He commercialized his invention and in the process initiated the electric generation industry. The new technology was quickly transferred to Europe and the rest of the world.

The rapid development of electric lighting threatened the oil industry, and in particular Standard Oil, with its massive investment in production, refineries, pipelines, storage and distribution facilities. The oil industry then welcomed the market that emerged for the newly created motor car, the first one created by Carl Benz in Germany. Another investor was the man who named the Ford after himself, namely Henry Ford, after resigning his position as chief engineer of Edison Illuminating Company in Detroit.

The automobile became the icon of the modern age. One writer stated, "The man who owns a motor car gets for himself, besides the joys of touring, the adulation of the walking crowd, and . . . is a god to the women." This confirmed how prescient Nahum (Chapter 2:4) was in the Old Testament when he stated: "The chariots shall rage in the streets, they shall jostle another in the broad ways: they shall seem like torches, they shall run like the lightning." The automobile would change the course of transport history and encourage the use of oil. As the author E.B. White stated: "Everything in life is somewhere else, and you get there by car."

A Global Industry

On January 10, 1901, in Beaumont, Texas, the discovery of the first real oil gusher, called Spindletop, launched the Texas oil boom. Sir Marcus Samuel, a political figure in London, set up a company in Texas called Shell Transport and Trading, named in honour of his father's early business in seashells.[5] Another major oil company that grew out of the Spindletop discovery was Texaco. Joseph Culliman established the Texas Fuel Company in 1897 for crude oil processing and marketing. It was a conservative but profitable company from the beginning.

The Spindletop find resulted in a significant increase in tankers to transport oil to Europe and elsewhere. Through the efforts of the Mellon family, Andrew, William and Richard, the Gulf Oil Corporation grew to control much of the early Texas production in conjunction with Shell. Shell became the second largest oil company in the world, while Standard Oil remained the largest. Shell amalgamated in 1907 with Royal Dutch Petroleum Company,[6] which was operating in Burma, Borneo and Sumatra. The amalgamated company was led by Henri Deterding who managed to unite with his great rival, Marcus Samuel of Shell Transport. Deterding and Samuel controlled more than half the Russian and Far Eastern oil exports.[7]

By the 1940s, a number of large North American companies had emerged in addition to Shell and the then Standard Oil of New Jersey (later to become Esso and Exxon): the new companies included Phillips, Mobil, Sun, Atlantic Richfield and Union. However, it was in Eastern Europe that the collision between politics and petroleum was the most dramatic. Before the First World War (1914–1918), Russian oil had been one of the most important elements in the world market. In 1918, that rich source of oil was firmly in the hands of the new Communist government of the Soviet Union. Just before the war, Royal Dutch/Shell had bought Rothschild's large oil interests in Russia. The Nobel family, led by Ludwig Nobel of Sweden, had developed profitable refining facilities at Baku. Ludwig Nobel in 1878 put into service the first successful bulk

tanker, a revolution in oil transport. During the Russian Revolution, which took place between 1918 and 1922, the Nobel family fled to Paris where they offered their properties for sale. Their properties, like the properties of others, were appropriated by the Soviet Government and the foreign companies therefore lost their investment.

The next major event in international oil occurred in the Middle East. Oil was discovered in Bahrain in 1932, and in Kuwait and Saudi Arabia in 1938. Saudi Arabia was on the road to immense wealth and the Kingdom's economy would no longer be vulnerable to fluctuations in the number of the faithful who made the pilgrimage to Mecca. The oil discoveries in the Middle East also resulted in many organizations and individuals making significant efforts to obtain concessions, which included the Iraq Petroleum Company (of Britain) and German, Italian, and Japanese interests.[8]

During the Second World War, new strategies were devised to produce more oil for war machinery. The person responsible for introducing geophysics into oil exploration, Everette DeGolyner, arrived in Saudi Arabia in 1943. His expertise was responsible for the new and very large oil finds which would allow Saudi Arabia to become the world's leading producer. The United States, no longer energy dependent at the end of the war, became susceptible to the vagaries of the Middle East and its largest company, Aramco. By this time the Saudis had begun to demand more money for their oil as had Venezuela, Iran and Kuwait.

The transport of oil spawned a new shipping industry in the mid twentieth century. The oil tanker business grew significantly in Europe, the Mediterranean, the United States and Japan. The Suez Canal became a major route-way for tanker traffic from the Middle East, boosting Britain's coffers but not that of Egypt, where the Canal was located. A dynamic Egyptian nationalist leader, Colonel Jamal Abdel Nasser, ordered his army to seize control of the Canal Zone in a single daring act on July 26, 1956. General Eisenhower of the US did not support the military intervention of Britain and France, fearing that an intervention might trigger an embargo on oil shipments from the entire Middle East. Fearful of the possible prolonged effect of a shortage of oil in Europe, the British withdrew from Egypt, an act which signalled the decline of Britain as a superpower. Little did Britain know that by 1970 large oil deposits would be discovered in its own North Sea. Had Britain been aware of its own oil resources, it would not have become involved in the Suez Crisis. Interestingly, the Suez Crisis also generated a new industry, that of supertankers. Supertankers became necessary, arising from the vulnerability of the Canal and Middle Eastern pipelines. Because the safer alternative was the route around the Cape of Good Hope, economic efficiencies required that much larger tankers transport much more oil. Soon after 1956, Japanese shipyards were building large numbers of these supertankers which would change the pattern of future worldwide oil transportation.

During the 1950s and 1960s, the battle for production intensified between Iran and Saudi Arabia. During 1957 and 1970, Iranian production grew faster than Saudi output. Consumers around the world welcomed cheap oil from the Middle East. The USA halted cheap oil imports in order to prevent the undercutting of the domestic industry and imposed import quotas. Thus prices were higher in the United States than they would have been without protection. In 1968, the price of oil at the wellhead in the US was $2.94 per barrel, some 60 per cent more than the Middle East crude oils fetched on the east coast of the US (Yergin, 1991).

The total world energy consumption began to explode. Demand tripled between 1949 and 1972. In Europe, demand increased fifteen times during that period, and in Japan one hundred times. The surge in oil use resulted from rising incomes, rapid economic growth and the fact that the low price of oil encouraged greater consumption. Coal was deposed as the king of fuels as oil became cheaper than coal. In search of higher revenues, the exporting countries sought to increase the volume sold, rather than increasing prices. Oil was searching desperately for markets and discounting was rampant. Discounting led to a split in the world oil industry between the 'posted' or official price, which was held constant, and the actual market price at which crude oil was sold, which at this time was falling. A country's taxes and royalties were calculated on the posted price which should closely match the market price. Discounting led to a significant gap between supply and demand but the suppliers would not drop the posted price because of the important revenues it provided. By 1960, the posted price had become a fictitious price, used only as a basis for computing revenues. Therefore the governments of the producing countries were taking a higher percentage of the profit made from the actual price realized by the companies. Simply put, the governments of producing countries were making good whilst the companies absorbed all the negative effects of the price cuts.

The market was tight in 1958 when the US imposed import quotas and significantly diminished the largest oil markets in the world. The situation was exacerbated by the re-entry of cheap oil from the Soviet Union into the market. Between 1955 and 1960, Soviet oil production doubled and the Soviet Union aggressively resumed exports to the West and also cut prices. The competitive response from the western companies was further price cuts. The western companies subsequently faced a monumental problem. By reducing the market price alone, the discount would have to be absorbed entirely by the companies. The western companies, led by British Petroleum in early 1959, made a ten per cent reduction on the posted price. This action led to a reduction of the national revenues of all oil-producing countries and the oil-producing countries had to take action.

At the 1959 Arab Oil Congress in Cairo, Abdullah Tariki, the Oil Minister of Saudi Arabia, and Perez Alfonso, the Oil Minister of Venezuela, formed an alliance that was to change the international dynamics of the petroleum industry. This alliance galvanized the idea of an oil cartel that would result in the formation of the Organization of Petroleum Exporting Countries (OPEC).

OPEC

OPEC's goal was to defend the price of oil by regulating production. The five founding members, Iran, Iraq, Kuwait, Saudi Arabia and Venezuela, controlled 80 per cent of the world's oil.[9] Almost as soon as OPEC had been formed, its member countries lost their virtually total grip on world oil exports. New oil provinces were developed and put on stream. Most of the new producing countries would eventually become members of OPEC, but they first entered the world market as competitors. Libya, for example, where small 'independents' produced most of the oil, flooded Europe with cheap oil in the first half of the 1960s. OPEC could claim two achievements in its early years. It ensured that the oil companies would be cautious about taking any major steps unilaterally and without consultation. It also ensured that the oil companies would not dare cut the posted price again.

In 1973, OPEC sought to institute a new international economic order by raising oil prices significantly and redistributing economic and political power. The member countries of OPEC thereby gained an important influence on the foreign policies of some of the most powerful countries in the world. There were major changes in the international oil market between 1973 and 1994, partly as a result of OPEC's ability to manipulate production. Before 1973, oil price forecasting had not been necessary because price changes were measured in cents and not dollars, a situation which would change dramatically in October 1973.

This dramatic change in October 1973 was due to a number of significant events. The momentum for price increases had been initiated by Libya's success in obtaining higher prices and taxes from the foreign oil companies operating there in 1970. Libya's success set the stage for the important Teheran Agreement of February 1971. The Teheran negotiations improved the collective bargaining power of OPEC. There were further price increases in January 1972 and June 1973 to compensate oil-producing nations for loss of income due to the devaluation of the US dollar. In late 1973 however, OPEC viewed the oil market as increasingly tight and OPEC members were pressured to produce more crude. In addition, inflation continued to erode sales revenues.

In October 1973, OPEC members began negotiations to obtain an increase in price with a two-fold justification. Firstly, the increase would compensate for inflation and secondly, would allow OPEC to extract a percentage of the oil companies' increasing profits. When the oil companies no longer cooperated in the negotiations, OPEC unilaterally raised the posted price of oil by US$2.00 per barrel. At the same time, the Arab oil-producing states imposed an embargo on oil exports to the USA and the Netherlands because of their support for Israel during the ongoing Arab–Israeli war. The Arab producers also reduced production by 20 per cent and promised further cuts until the USA stopped its support for Israel. The market shortages forced prices upwards as countries and private companies without their own oil reserves competed for supplies. In December, the posted price of oil was raised to US$11.65 per barrel. This increase raised the posted price to 300 per cent above pre-1973 prices and a new plateau of high oil prices had been reached.

During 1974 to 1978, there were no volatile increases in oil prices. The constraint on price increases was brought about mainly by Saudi Arabia's reluctance to reduce its output. Saudi Arabia's premise was that higher oil prices would lead industrial consumer countries to develop and utilize alternative energy technologies and therefore ultimately reduce the market for Saudi Arabia's giant oil production potential. In early 1979, the upheavals associated with the Islamic Revolution in Iran led to a severe decline in that country's oil production. Prices increased to ensure control demand. By April, Iranian oil production resumed albeit at a lower output level, but prices continued to slowly move upward throughout the remainder of the year. In September 1980, OPEC reached an agreement on the official price of US$30 per barrel for marker crude. But soon the oil market experienced another supply disruption when war broke out between Iran and Iraq. The decline in oil output was offset by production increases in other OPEC countries and by a gradual reduction in demand due to conservation in the oil-deficient countries. The OPEC marker crude price was raised to US$34 per barrel in January 1982 when the spot market price was approximately US$36 per barrel.

The growth in non-OPEC production, in addition to the draw-down of stockpiles, lower economic growth, and demand reduction by oil-deficient countries, resulted in softer market conditions in 1982 and 1983. OPEC was able to maintain its marker crude price until March 1983 by lowering its output. For example, OPEC production in 1979 was 31 million barrels per day (b/d) compared to 17 million b/d in 1983. In March 1983, OPEC decreased its official price to US$29 per barrel, set an output limit of 17.5 million b/d and gave a quota to each member. Saudi Arabia was not given a quota, however, but was left to determine its own production level. Hence, it became the 'swing' producer, ensuring that oil prices were balanced by supply and demand.

Increasing non-OPEC production, coupled with conservation measures that had reduced world demand, led to a new agreement on production ceilings on October 30, 1984. Production was to be cut by 1.5 million b/d making the new output ceiling 16 million b/d. The agreed limits were not adhered to by OPEC members and this led to large cuts in Saudi Arabia's production during 1985. Saudi Arabia abandoned its uncomfortable role as 'swing' producer and in November 1985 increased its output from 2.3 million b/d to 4.2 million b/d, with a further rise in production to 6.4 million b/d in August 1986. A soft market, in tandem with a significant increase in output, brought about a dramatic decline in oil prices in July 1986. Prices shrank even lower than US$10 per barrel on the spot market.

In August 1986, oil prices experienced a modest rebound due to two factors: a return to OPEC production ceilings set in 1983 and an increased demand attributable to economic growth. The new price for marker crude was US$18 per barrel. The outcome of the 1986 price war was a decrease in the price of crude oil to approximately US$9.00 in 1987. In real terms, this price was lower than the price in 1974. The declining trend in real oil prices over the period 1980 to 1987 was steep, and made alternative energy sources less competitive during the mid-1980s.

OPEC and the volatile politics of the Gulf region had a powerful effect on the oil market. Oil and politics are inextricably linked in the Middle East so that political and strategic considerations have often intruded into the world of oil. The former concessionary system of exploration and production gave oil companies too much power over production and pricing, which eventually led to confrontations with host governments.

Prices and Production

In July 2008, oil prices reached US$147 per barrel. The effect of a recession which began to play out in September of the same year saw a significant reduction in oil prices which reached lows of US$32 dollars per barrel in January 2009. It is expected that oil prices will make an upward movement as the world economy grows out of the recession and will once again move towards the US$100 per barrel mark and beyond. Emphasis should now be placed on increased exploration and production, especially in the western hemisphere and in Russia and the oil-rich former states of the Soviet Union. This emphasis will help to adjust the present imbalance where nearly 60 per cent of world oil reserves lie beneath Middle East sands. As shown in Figure 2.1, the Middle East is a major source of oil for export around the world.

World oil prices are expected to continue to rise slowly over the next decade. The market's present calm reflects a belief that world production capabilities are sufficient to move the market through all but the worst-case scenarios incurred by wars or political crises. Oil prices may stabilize to respond to increased demand, especially from the Asia–Pacific region.

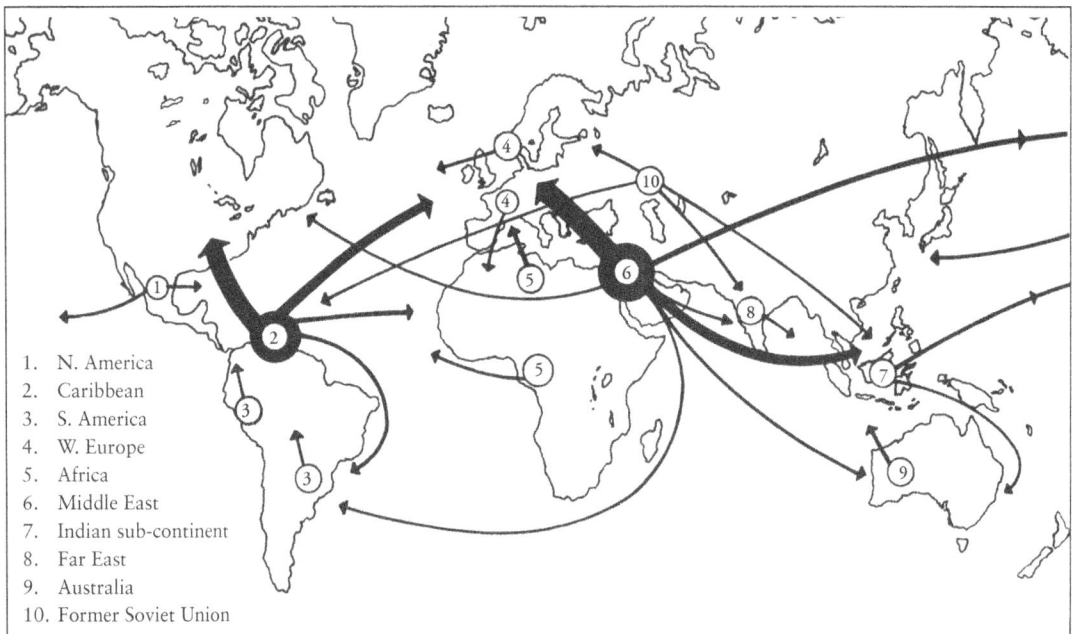

1. N. America
2. Caribbean
3. S. America
4. W. Europe
5. Africa
6. Middle East
7. Indian sub-continent
8. Far East
9. Australia
10. Former Soviet Union

Figure 2.1 International Movements from Major Oil-Producing Countries

Of course, there will be fluctuations influenced by political swings or wars. Energy conservation will help to restrain future oil price increases, although by the year 2020 the non-renewable nature of oil may encourage gradual price increases.

No large change in production quotas appears likely in the near future, although the world demand for oil is rising gradually and is likely to reach approximately one million barrels per day or more. The relatively low price of oil is encouraging this level of demand while industrial expansion in emerging nations, China and India for example, has created an increase in consumption.

OPEC's share of world oil production is expected to expand due to lower output in the United States. Notably, there are projects occurring in other regions of the world. One of these projects would link Azerbaijan with Russia's southern pipeline system for transport to Novorossyisk on the Black Sea, where there is a terminal that would be expanded. The other proposed project is to transport oil from Kazakhstan and Azerbaijan through Georgia and Turkey to a terminal at Ceyhan on the Mediterranean. The third proposal is to route the oil through Iran, taking the Caspian crude into the Persian Gulf with access to Asia. During 1996 to 1998 decisions on these matters will be taken either unilaterally or cooperatively.

Over the next 25 years, OPEC envisages that production from its 12 member states will increase to approximately 50 million b/d, an optimistic estimate, given the many imponderables that impact oil production and prices. Many analysts agree that the influence of OPEC could remain if it avoids the internal rifts of the past. There is however, a looming threat to OPEC's survival presented by both natural gas and the increase in non-OPEC oil. Natural gas is an emerging threat which could usurp oil as the world's primary energy supply. The ultimate return of Iraq to the export market, after having been barred by the UN from exporting oil since its invasion of Kuwait in 1990, as a factor has had an effect on oil production and pricing.

Non-Conventional Petroleum: Oil Shales

Oil shales are a potential source of non-conventional petroleum. Oil shales are found in many regions in the world, but the largest deposits are found in the United States in the Green River Formation, which covers sections of Colorado, Utah, and Wyoming. The estimates of the oil resources within the Green River Formation range from 1.2 to 1.8 trillion barrels. Not all resources in place are recoverable; nonetheless, even an optimistic estimate of 800 billion barrels of recoverable oil from oil shale in the Green River Formation is nearly three times more than the proven oil reserves of Saudi Arabia.

Oil shale can be mined and processed to generate oil similar to oil pumped from conventional oil wells; however, extracting oil from oil shale is more complex than conventional oil recovery and is currently more expensive. The oil substances in oil shale are solid and cannot be pumped directly out of the ground. The oil shale must be mined, heated to a high temperature (a process called retorting), and the resultant liquid separated and collected.

While current technologies are adequate for oil shale mining, the technology for surface retorting has not been successfully applied at a commercially viable level in the United States, although technical viability has been demonstrated. Further development and testing of surface retorting technology is needed before the method is likely to succeed on a commercial scale.

Oil Deposits Outside of the Middle East, Europe and the United States

There are massive oil deposits outside of the Middle East, Europe and the United States. Located in northern Alberta, Canada, are the Athabasca Oil Sands, a large deposit of oil-rich bitumen, or extremely heavy crude oil. These oil sands consist of a mixture of crude bitumen (a semi-solid form of crude oil), silica sand, clay minerals, and water. The Athabasca deposit is the largest of three major oil sands deposits in Alberta, along with the nearby Peace River and Cold Lake deposits. Together, these oil sand deposits cover about 141,000 square kilometres (54,000 sq mi) of sparsely populated boreal forest and muskeg (peat bogs), contain about 1.7 trillion barrels (270 × 109 m^3) of bitumen in-place, and is equivalent to more than 50 per cent of the world's total proven reserves of conventional petroleum. With present technology, about 10 per cent of these deposits, or about 170 billion barrels (27 × 109 m^3), are considered to be economically recoverable at prices above US$60 per barrel, giving Canada oil reserves second in the world only to Saudi Arabia. The key characteristic of the Athabasca deposit is that it is the only one shallow enough to be suitable for surface mining. About 10 per cent of the Athabasca oil sands are covered by less than 75 metres (246 ft) of overburden. The mineable area as defined by the Alberta government covers 37 contiguous townships (about 3,400 square kilometres (1,300 sq mi) north of the city of Fort McMurray. The overburden consists of 1 to 3 metres of water-logged muskeg on top of 0 to 75 metres of clay and barren sand, while the underlying oil sands are typically 40 to 60 metres thick and sit on top of relatively flat limestone rock.

With planned projects coming on stream, by 2010 oil sands production is projected to reach 2 million barrels per day (320,000 m³/d) or about two-thirds of Canadian production. By 2015 Canadian oil production may reach 4 million barrels per day (640,000 m³/d), of which only 15 per cent will be conventional crude oil.

The Venezuelan Orinoco tar sands site may contain more oil sands than Athabasca. However, while the Orinoco deposits are less viscous and more easily produced using conventional techniques (the Venezuelan government prefers to call them "extra-heavy oil"), they are too deep to access by surface mining.

Future Trends

The primary energy needs of the world will continue to be met by oil well into the second half of the twenty-first century. However, over the next decade OPEC may continue to experience difficulties reaching its price objectives because the need to agree on fixed outputs may be outweighed by the incentive to produce secretly beyond agreed ceilings. There are two other factors which may constrain the wide power of OPEC at least in the near future: the lack of stringent enforcement mechanisms within OPEC, and the fact that about 60 per cent of the world's crude oil production is now outside of OPEC. These circumstances are known to OPEC which intends to promote more controls and efficiency in its production activities. The oil revenues of OPEC countries have sustained comparatively strong internal economies when compared with oil-deficient nations. Figure 2.2 shows the world reserves for oil and indicates the high percentage of reserves in OPEC countries and Saudi Arabia in particular.

Energy substitutes for oil will help to limit future increases in oil production. Some major oil companies are investing in renewables but they are remaining solely in the energy sector, unlike the situation in the 1970s

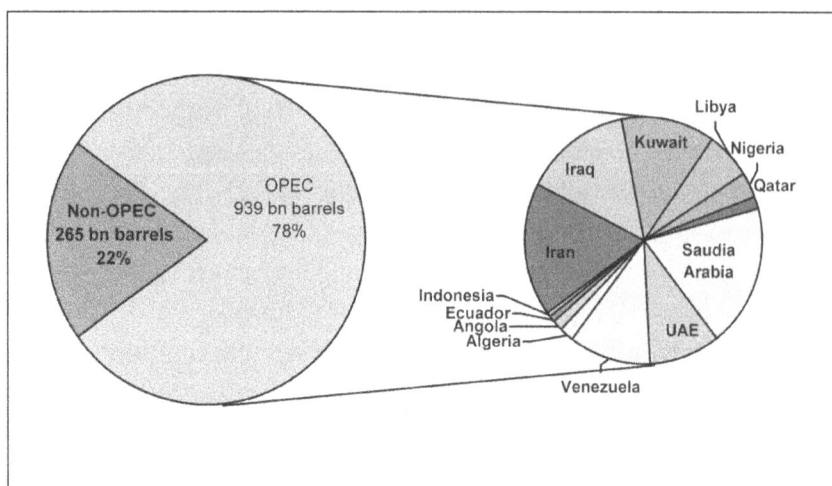

Figure 2.2 OPEC Share of World Crude Oil Reserves (2007)
Source: OPEC

and early 1980s when oil industries strayed into other industries such as real estate development. This focus will facilitate the strengthening of the oil industry. The oil industry is also resisting the influence of invalid theories, notably the pessimistic forecasts on depleting resources. Companies are strategically making long-term investments in both upstream and downstream activities. Many oil companies are diversifying into other energy resources, for example renewables, and EN CON and others are becoming independent power producers. At the same time, energy strategies are becoming globalized and oil in the former states of the Soviet Union is now an integral part of the international energy economy.

OPEC will become a less important force in the oil supply market as non-OPEC producers such as the Russian Federation, members of the former Soviet Union and emerging nations such as Brazil ramp up production. An interesting development is that more of the world's oil resources are being marshalled and managed by governments through national oil companies. As a consequence, the large multinational oil companies have had their access to resources and resource potential reduced. The multinationals have not been able to spend the large fiscal requirements for effecting technological developments in the oil fields of many of these countries. Hence, there has not been a significant increase in supply due to inadequate investment in new exploration ventures. In this scenario, Saudi Arabia continues to exercise its role as a swing producer being the only country capable of increasing production with alacrity and immediacy. The production in some OPEC countries, such as Indonesia, has been reduced significantly. Thus global oil resources will see a gradual shift towards production from non-OPEC sources. There will be a new thrust in searching for oil in new provinces in the Caribbean, Africa and the Arctic Circle.

Future trends will also be guided by the huge global political impact of oil. Over almost a century and a half, oil has been both boon and burden. As Yergin (1991) notes, of all energy sources, it has loomed the largest and the most problematic because of its central role, strategic position, geographic distribution and crisis in supply. We are now being challenged yet again by a political, economic and environmental oil crisis. Dedication, ingenuity, entrepreneurship and innovation have co-existed with avarice, corruption, political ambition and military force. Oil prices have increased dramatically on a tide of influences relating to supply and demand, geopolitical problems, speculation as well as the decline in the value of the US dollar as a global currency. It is difficult to predict when global oil supplies will peak. Among the factors affecting peak oil, as Holland (2008) posits, are increases in production, increasing demand, cost reductions through technological change, cost reductions through exploration and increasing production from additional site exploration and development. It is expected that oil will remain as an important primary fuel for another 50 years but will continue to increase in value and in price. Oil will eventually become too valuable a commodity to be used for electricity generation and we will gradually see its use being almost exclusively directed towards the transportation and petrochemical industries.

The future of the oil market will depend in large measure on the fundamental balance between supply and demand which is required to avoid speculation by traders using oil as just another profit making commodity. John Lipsy of the IMF in a May 2009 presentation to the G8 meeting of Energy Ministers in Rome made a pertinent observation which succinctly combines the major issues facing the oil and gas future. He stated that "supply constraints eventually could reemerge, threatening medium-term oil market stability. To avoid excessive price swings, efforts by producers, consumers, financial investors, and market regulators alike will be needed to improve the transparency, functioning, oversight and, ultimately the supply-demand balance, in global oil markets."

There has been an adverse effect on the automobile industry from the global economic recession of 2009. Large automobile manufacturers such as General Motors and Chrysler have been downsized and there has been a reduction in the rate of demand for motor vehicles. This demand may move upward as the world emerges from the recession. Technological trends are moving in the direction of hybrid vehicles, electric cars, Compressed Natural Gas (CNG) and LPG as automobile fuel as well as hydrogen fueled cars. These trends point to a gradual reduction in the demand for oil. However, this expected lowering of per capita demand, if it is sustainable, will be reflected gradually over the longer term. In the meantime, oil continues to be the most sought after and widely distributed fuel source. Its future utilization will require policy directives and proper management based on best efforts to develop new producing fields at the lowest cost. The reduction in speculation in the market place, and a better understanding of the supply and demand requirements globally are also essential to price stability and oil availability.

References

Aeseng, N. (2000) *Business Builders in Oil Minneapolis*. The Oliver Press.

Crow, P. (1996) "Iran Seeks to be Outlet for Caspian Sea oil". *Oil and Gas Journal* 94 (1) pp. 29–30.

Holland, S. (2008) "Modeling Check Up Peak Oil". *The Energy Journal* 29 (2) pp. 61–79.

Lipsky, J. (2009) "Prospects for Oil Market Stability". Remarks at the G-8 Meeting of Energy Ministers Rome, May 25.

Lynch, M. (1994) "Shoulder Against Shoulder: the Evolution of Oil Industry Strategy". Working paper 2609. MIT: Centre for International Studies.

Yergin, D. (1991) *The Prize: the Epic Quest for Oil, Money and Power*. New York: Simon & Schuster.

Yergin, D. and Gustafson, T. (1995) *Russia 2010, and What it Means to the World*. New York: Vintage Books.

3. | Searching For New Petroleum Resources in a Frontier Province

The Case of Jamaica

As STATED IN THE PROLOGUE, Jamaica is highlighted as an illustration of an energy deficient country which is heavily dependent on imported oil. Most of the foreign exchange earned by the export of merchandise in Jamaica is used to purchase imported fuel. There have been important efforts to find oil and gas in Jamaica, but the discovery and extraction of oil is not a simple process.

Oil and gas originated globally from the decomposition of aquatic, primarily marine, plants and animals buried beneath successive layers of mud and silt in geologic time some 60–500 million years ago. Conditions on the sea bottom were such that no rapid decay of the dead organisms by bacterial action could take place. Over time, mud and silt layers deposited upon the potential source beds produced pressures and higher temperatures in the beds. As the source rocks became compacted by over-burden pressure, oil and gas, with associated water, were squeezed out. The fluids migrated into porous formations and accumulated when trapped by impermeable conditions. The impermeable layer that prevents further movement of fluid is a 'seal' or 'caprock'.

Conditions suitable for the generation and accumulation of oil exist in the down-warped segments of the earth's crust where layers of sediment have piled up to great depths, thickest in the middle and thinner towards the edges. Oil may be generated in sedimentary rocks of almost any age, but the older rocks have lost more of their oil than the younger ones through seepage, erosion or heat from igneous and volcanic rocks.

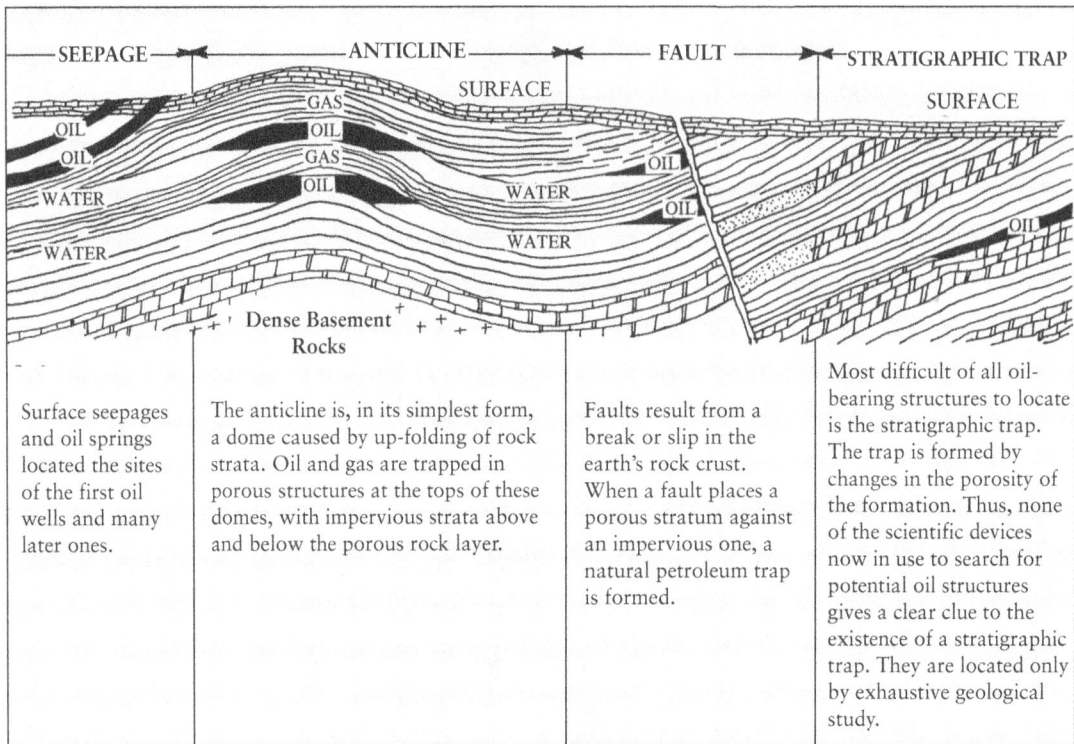

SEEPAGE	ANTICLINE	FAULT	STRATIGRAPHIC TRAP
Surface seepages and oil springs located the sites of the first oil wells and many later ones.	The anticline is, in its simplest form, a dome caused by up-folding of rock strata. Oil and gas are trapped in porous structures at the tops of these domes, with impervious strata above and below the porous rock layer.	Faults result from a break or slip in the earth's rock crust. When a fault places a porous stratum against an impervious one, a natural petroleum trap is formed.	Most difficult of all oil-bearing structures to locate is the stratigraphic trap. The trap is formed by changes in the porosity of the formation. Thus, none of the scientific devices now in use to search for potential oil structures gives a clear clue to the existence of a stratigraphic trap. They are located only by exhaustive geological study.

Figure 3.1   Types of Trapping Mechanisms for Oil and Gas

Oil traps are of many types, broadly falling into structural and strati-graphic traps. Structural traps result from some deformation such as fold-ing, faulting, or both, of the reservoir and seal. Changes in the lithologic composition of sedimentary rocks may cause a lateral decrease in permeability and porosity. The more porous section of the layer may form a stratigraphic trap, which accumulates oil as a result of changes in surrounding rock permeability characteristics (Figure 3.1). Such traps are often formed by the updip edge of a wedging-out sand layer that merges into an impervious clay of a lenticular nature and enclosed in tight sedi-ments such as shales.

The formation and accumulation of oil and gas therefore require a number of factors to be in place, the most important of which are:

- organic source materials, exposed to the pressure and temperature necessary to convert them into petroleum
- carrier beds to convey the petroleum to an accumulation site
- sealed traps large enough to enable the accumulation of commercial volumes of petroleum

- preservation of the traps over geologic time to prevent the seals from being broken by erosion. This erosion can be triggered by tectonic activity such as uplift, folding and faulting

The absence of any one of these requirements on a large scale makes it unlikely for petroleum to accumulate in quantities sufficient for commercial use. Some of these factors obtain in Jamaica. Of importance, the accumulation of oil in the Jamaican province, either onshore or offshore, can occur in large quantities provided there are sufficient source rocks and sealing mechanisms that will trap the oil.

In order to reach the oil and gas accumulated in the traps, it is necessary to drill through the impervious layer which would have performed as a seal. Drilling is conducted by a rotary drilling installation consisting of bit, drilling string, rotating equipment, hoisting equipment, and drilling fluid circulating equipment and engines (Figure 3.2). Figure 3.3 is a diagram of the seismic geophysical process and Figure 3.4 is an illustration of a pumping well.

The areas of knowledge used in petroleum exploration include geology, geochemistry and geophysics, all of which form part of a pre-drilling exploration programme. Presently, the only way to know if oil is in the ground is to drill, which is very costly, but fortunately, various remote sensing techniques are now being developed. A 4,000m well drilled onshore Jamaica could cost approximately US$12 million; a well drilled to the same depth offshore could cost upward of US$36 million depending on water depth. Therefore, when a hole is drilled and no oil is found, considerable sums of money are lost.

A typical onshore gas seep. This example is from Windsor St Ann Jamaica

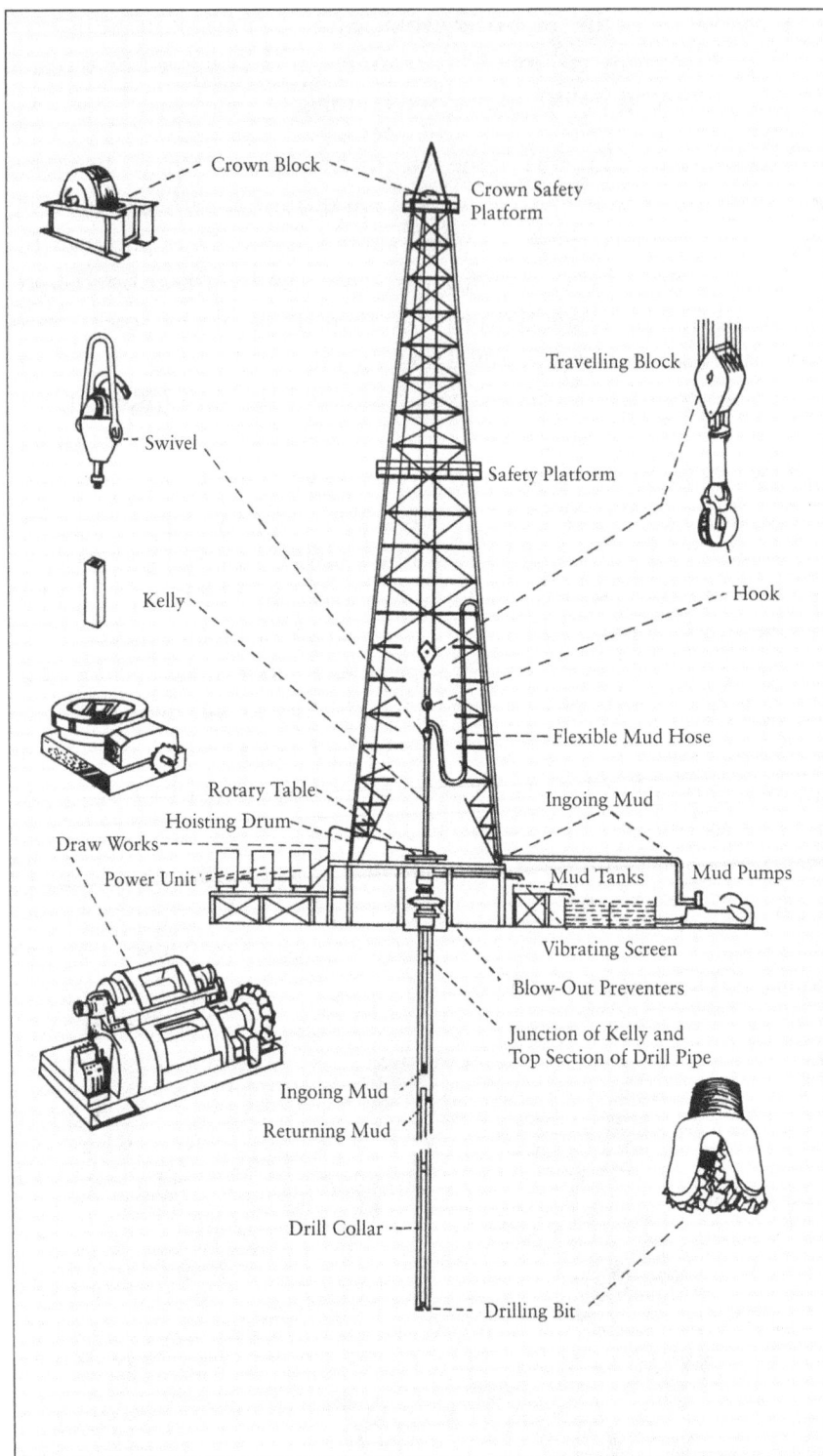

Figure 3.2 Simplified Diagram of a Rotary Drilling Rig (not to scale)

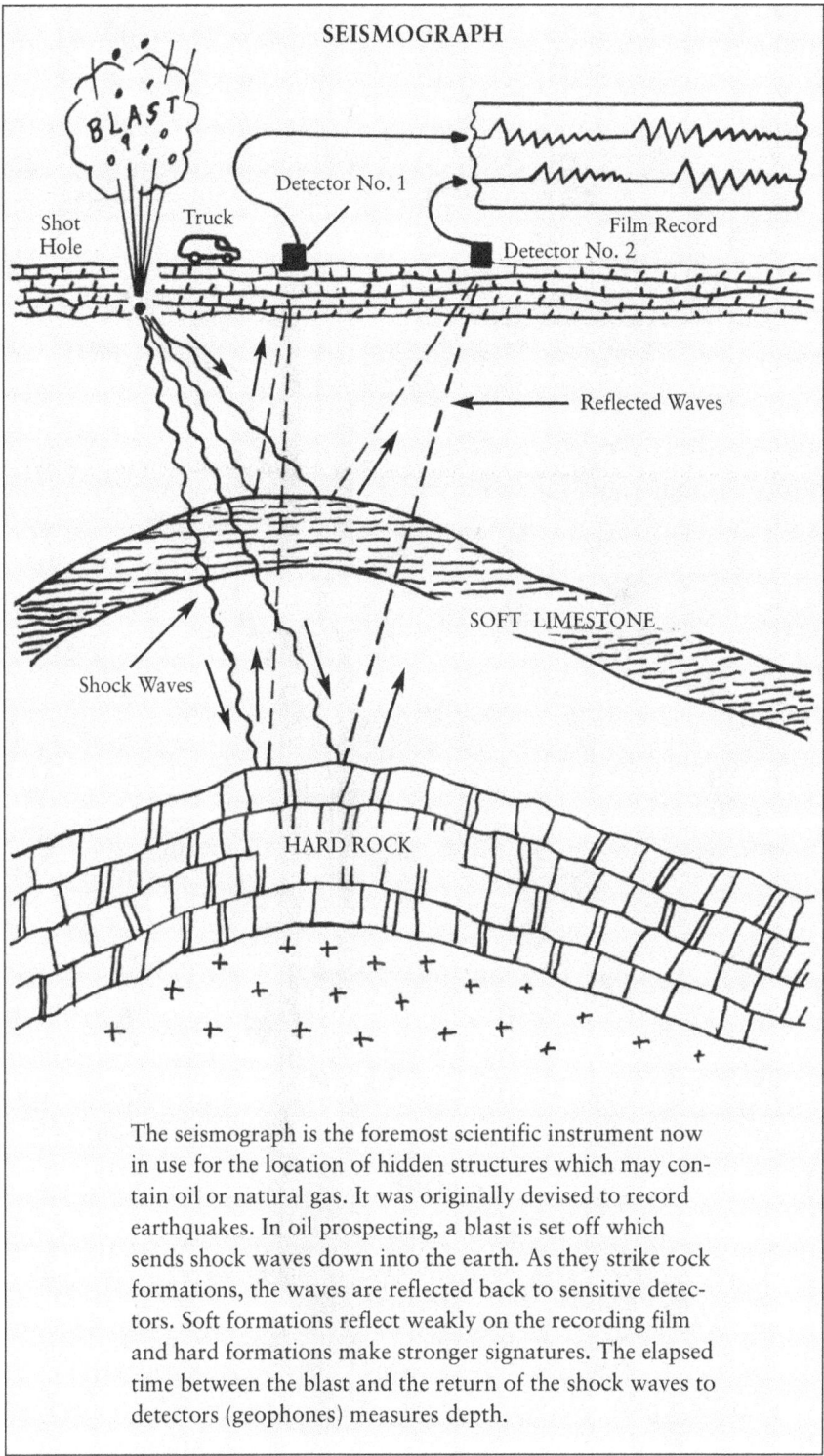

SEISMOGRAPH

BLAST

Shot
Hole

Truck

Detector No. 1

Detector No. 2

Film Record

Reflected Waves

Shock Waves

SOFT LIMESTONE

HARD ROCK

The seismograph is the foremost scientific instrument now
in use for the location of hidden structures which may contain oil or natural gas. It was originally devised to record
earthquakes. In oil prospecting, a blast is set off which
sends shock waves down into the earth. As they strike rock
formations, the waves are reflected back to sensitive detectors. Soft formations reflect weakly on the recording film
and hard formations make stronger signatures. The elapsed
time between the blast and the return of the shock waves to
detectors (geophones) measures depth.

Figure 3.3 Diagram of Seismic Geophysical Process

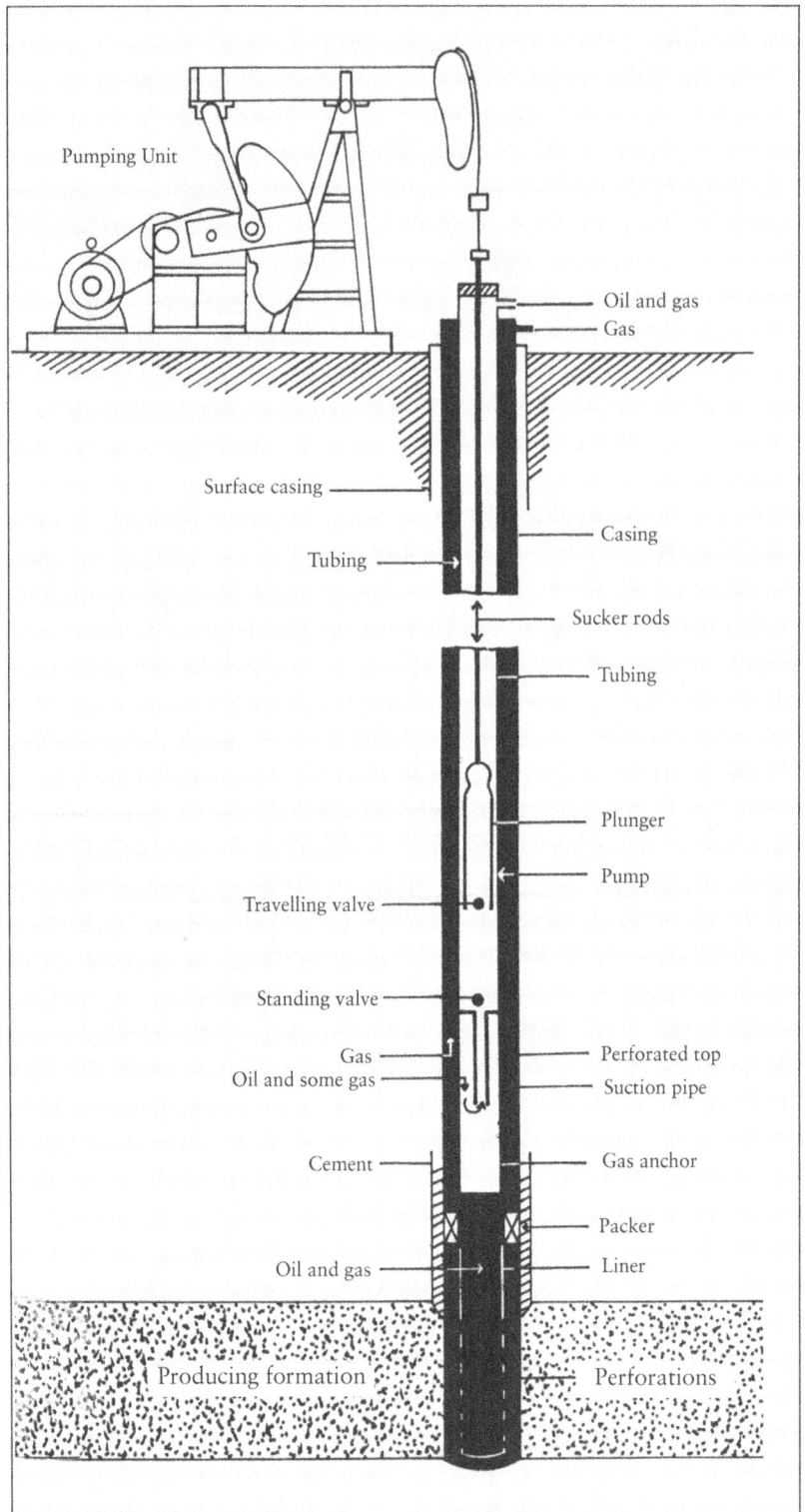

Figure 3.4   Diagram of a Pumping Well

Due to the potential loss of huge sums of money, it is better to have private companies rather than the state drill for oil in small nations. In the event of a dry hole, the private company can write off much of the expenditure against taxes, so that only a small part of each dollar is actually risked and lost to the company. When a state-owned company spends a dollar to drill, it costs the national treasury nearly 100 cents, especially if foreign contractors are used because there is no 'genuine' tax write-off, since both the state company and the government belong to the same entity.

When oil is found, the well will produce by natural flow, gas lifting or pumping. Most producing wells are operated by mechanical lifting methods using subsurface pumps.

The search for oil in Jamaica can be divided into two phases, the first involving private industry entirely, and the second primarily involving, the state-owned company, Petroleum Corporation of Jamaica (PCJ). The first phase spanned the period 1955–1973, whilst the second phase covers the period 1978 to the present. The exploratory operations have been conducted both onshore and offshore, the Pedro Banks being the main offshore target.

Eleven wells have been drilled in Jamaican territory (Figures 3.5, 3.6, Table 3.1). Between 1955 and 1973, seven exploratory wells were drilled, six onshore and one offshore on Pedro Bank. The onshore wells were

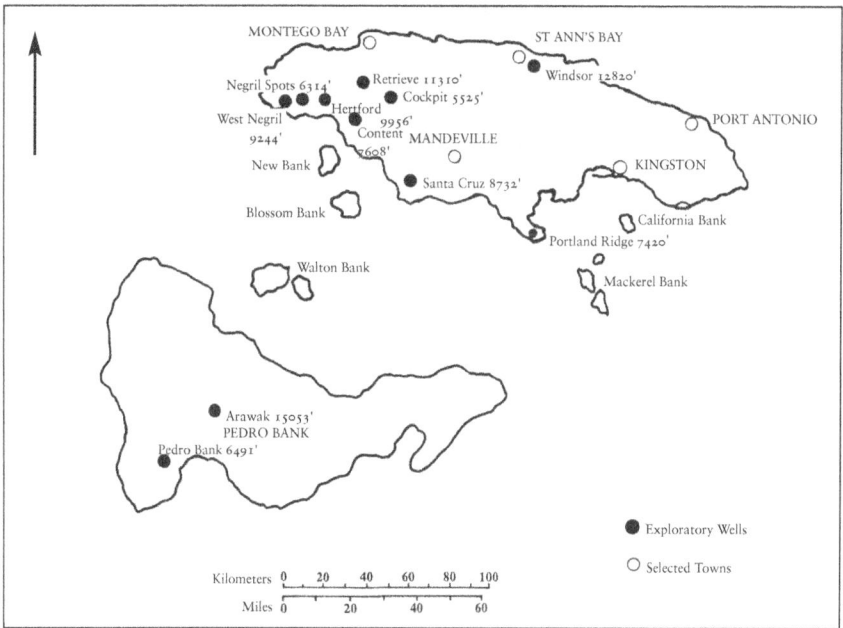

Figure 3.5 Location of Exploratory Wells in Jamaica

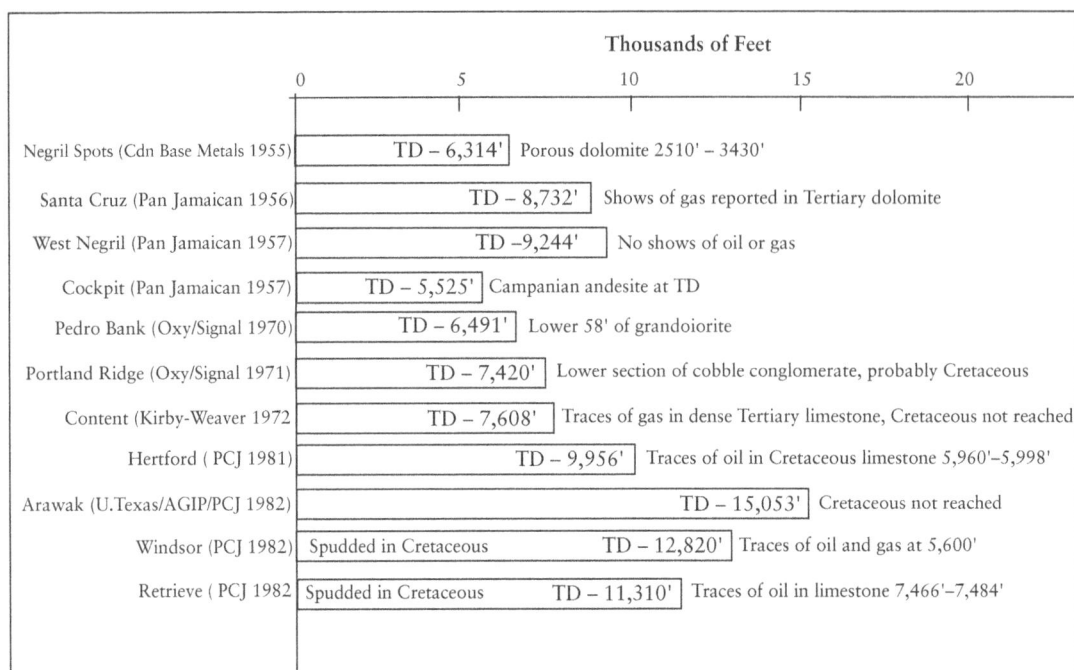

	Thousands of Feet				
	0	5	10	15	20

Negril Spots (Cdn Base Metals 1955) — TD – 6,314' — Porous dolomite 2510' – 3430'

Santa Cruz (Pan Jamaican 1956) — TD – 8,732' — Shows of gas reported in Tertiary dolomite

West Negril (Pan Jamaican 1957) — TD –9,244' — No shows of oil or gas

Cockpit (Pan Jamaican 1957) — TD – 5,525' — Campanian andesite at TD

Pedro Bank (Oxy/Signal 1970) — TD – 6,491' — Lower 58' of grandoiorite

Portland Ridge (Oxy/Signal 1971) — TD – 7,420' — Lower section of cobble conglomerate, probably Cretaceous

Content (Kirby-Weaver 1972) — TD – 7,608' — Traces of gas in dense Tertiary limestone, Cretaceous not reached

Hertford (PCJ 1981) — TD – 9,956' — Traces of oil in Cretaceous limestone 5,960'-5,998'

Arawak (U.Texas/AGIP/PCJ 1982) — TD – 15,053' — Cretaceous not reached

Windsor (PCJ 1982) — Spudded in Cretaceous — TD – 12,820' — Traces of oil and gas at 5,600'

Retrieve (PCJ 1982) — Spudded in Cretaceous — TD – 11,310' — Traces of oil in limestone 7,466'-7,484'

Figure 3.6 Past Exploratory Wells – Jamaica

drilled at Negril Spots (1955); near Munro College in the Santa Cruz Mountains (1956); in the southern part of the Cockpit Country (1957); West Negril (1957); Portland Ridge (1971); and Content, Westmoreland (1972).

The Negril Spots Well was drilled by Canadian Base Metals; and Santa Cruz # 1 (sited 2 kilometres SE of Munro College) together with West Negril and Cockpit (west of Troy) were drilled by Pan Jamaican. Occidental, in a joint venture with Signal, drilled the first offshore well on Pedro Bank and an onshore well at Portland Ridge.

A group led by O.D. Weaver drilled a well on an anticlinal structure at Content, east of Bluefields in Westmoreland in 1972. Their geologist, W.B. Harriman, was to become one of the great proponents of the petroleum prospectivity of onshore and offshore Jamaica.

During the period 1973–1976, only geological data gathering was conducted and this was achieved largely by Harriman, until an evaluation of existing data, spearheaded by A.A. Meyerhoff and E.A. Krieg in 1976–1977, stimulated further seismic geophysical work offshore and onshore, resulting in the acquisition of a comprehensive programme of seismic data in 1978–1979. That geophysical data acquisition was funded

Table 3.1.   Summary of exploration wells in Jamaica 1955–1982

Year	Onshore	Total Depth (ft)
1955	Negril Spots, drilled by Canadian Base Metals	6,314
1956	Santa Cruz, drilled by Pan Jamaican	8,732
1957	West Negril, drilled by Pan Jamaican	9,244
1957	Cockpit, drilled by Pan Jamaican	5,525
1971	Portland Ridge, drilled by Oxy-Signal	7,420
1972	Content, drilled by Kirby-Weaver	7,608
1981	Herford, drilled by PCJ	9,956
1982	Windsor, drilled by PCJ	12,820
1982	Retrieve, drilled by PCJ	11,310
Year	**Offshore**	
1970	Pedro Bank, drilled by Oxy/Signal	6,492
1982	Arawak, drilled by Union Texas/AGIP/PCJ	15,053

by the Government of Norway through its aid agency, NORAD. Meyer-hoff and Krieg (1977) prepared an important report which made a comprehensive analysis of the petroleum potential of Jamaica. The report was positive and encouraged exploration activities over the next eight years.

After the Petroleum Corporation of Jamaica had been formed in June 1979, the momentum of exploration activity increased. During the period 1981–1982, PCJ, with assistance from the Inter-American Development Bank, drilled three wells onshore at Hertford in Westmoreland, (9,956 ft), Retrieve, St James (11,310 ft), and Windsor, St Ann (12,820 ft). An offshore well on Pedro Bank was drilled by Union Texas/AGIP to a depth of 12,005 ft in 1981. It was subsequently deepened by PCJ, with assistance from the World Bank, to 15,053 ft. The most encouraging development involved oil and gas shows in two of the wells (Windsor and Retrieve) drilled by PCJ.

Knowledge of petroleum prospectivity was further enhanced during the 1980s by surface geological mapping, geochemical surveys and some seismic surveys. Much of this activity was supported by financial and technical assistance from the Canadian government through Petro-Canada International Assistance Corporation. The seismic surveys identified an attractive prospect on New Bank, offshore southern Jamaica, and enhanced our understanding of five potential oil-bearing structures onshore. In all, approximately 31,905 line km of seismic geophysical data have been collected offshore, and 948 km onshore.

The evaluation of the body of geological, geophysical and geochemical data indicates that a number of structures occur offshore and onshore which may warrant drilling. Offshore, the structures include one on New Bank, a number in the Walton Basin and on the northern side of Pedro Bank. Seismic work conducted in 2009 indicates a number of prospects including the deeper water areas. It is expected that the search for oil will follow these deeper water leads.

Onshore, there are smaller structures with teasing potential. These are mainly structural traps and include Ecclesdown (Portland); the northern end of the Santa Cruz Mountains and north of Middle Quarters (St Elizabeth); near Mount Airy, south of Negril (Westmoreland); east of Montego Bay (St James) and north of Bamboo (St Ann). The onshore structures are a mixture of anticlinal and stratigraphic traps.

One of the most important pieces of evidence relating to the generation of natural gas onshore Jamaica is a gas seep at Windsor in St Ann which was first recorded in 1896 and continues to flow at present. Other gas seeps have been recorded in the north coast belt near Annotto Bay.

In 2004, oil and gas activity was renewed by the preparation of a report by the consulting firm JEBCO. This report presented a detailed synthesis and analysis of Jamaica's oil and gas prospectivity with particular reference to offshore. It initiated the announcement of a formal bid round for the Jamaican acreage which was then divided into 24 blocks, 4 onshore and 20 offshore. Bids were received on July 15, 2005 following road shows in London, Houston and Cancun. Contracts were awarded to two companies for offshore exploration blocks. Blocks 6, 7, 10, 11 and 12 were awarded to a Joint Venture between Finder Exploration and Gippsland Offshore Petroleum of Perth, Australia. Blocks 9, 13 and 14 were awarded to Rainville Energy of Canada. In December 2007 following an informal bid round, four blocks, 1, 5, 8 and 17 were awarded to a Hong Kong based company, Proteam.[1] During 2006 to 2009, approximately 9,600 line kms of seismic data and 23,000 kms of gravity and magnetic data were collected offshore by two of these companies. In 2009, the Jamaican offshore area was subdivided into a total of 31 blocks (Figure 3.7) and new geophysical data, amounting to 6,217 line kms were acquired for the open acreage on a multi client basis. The open Jamaican

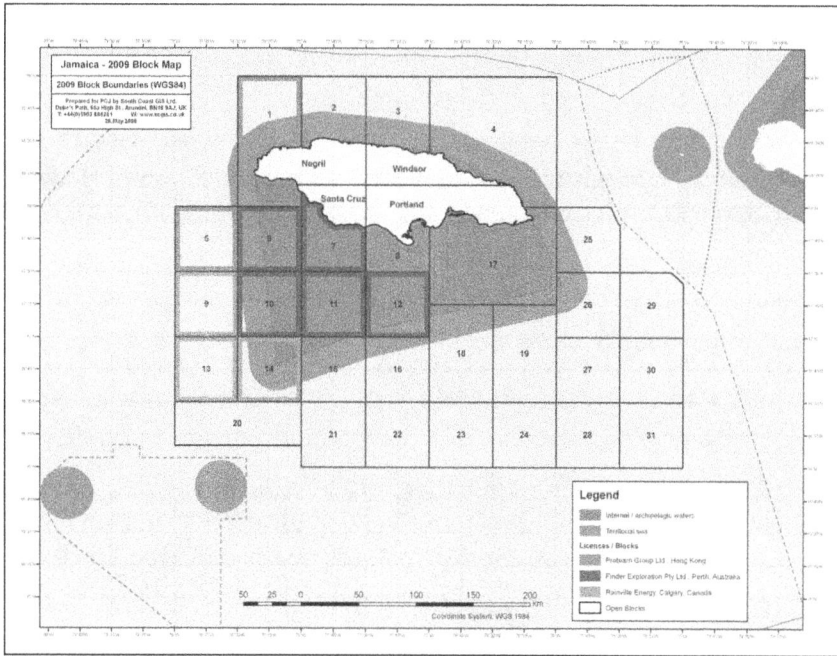

Figure 3.7   Jamaica Block Map

offshore acreage will be the subject of a second formal bid round in 2010. The offshore area, based in large measure on the new seismic data is highly prospective. Within the open acreage there are interesting features such as grabens, horsts, tilted fault blocks, carbonate build-ups, and large structures with thick sedimentary sequences in excess of 4,000 m.

The new geoscientific investigations have pointed to the fact that there could be three petroleum systems operating in Jamaica: one in the lower Tertiary carbonates, the second in the upper cretaceous shales and sandstones, and a third in lower cretaceous sandstones and other clastics.

Within the Walton Basin there are source kitchens as well as structures with potential reservoir capacity. Given the indications of mobile hydrocarbons, the possibility exists of large finds of both oil and gas. As reservoirs and seal studies continue, coupled with Basin modelling for source maturation and migration, the hydrocarbon potential of offshore southern Jamaica is being enhanced. Of course, only drilling will be able to prove whether commercial oil or gas exists. However, a number of remote sensing techniques are now being developed to prove the occurrence of oil prior to drilling. These techniques include:

- Remote Reservoir Resistivity Mapping (R3M), which is a sophisticated technology based on a simple fact: that oil and gas are poor conductors of electricity. R3M uses extremely low-frequency electromagnetic waves to discern these resistive deposits – remotely "mapping" undersea oil and gas reservoirs with some accuracy.

- The Sedimentary Residual Magnetic (SRM), which uses an airborne cesium vapor magnetometer to detect any accumulation or body of magnetic material having higher magnetic susceptibility than the surrounding sediments.
- Magnetic Bright Spot (MBS), which is defined as the area within the outline around the clustering of line to line SRM anomalies.

A Joint Regime Area (JRA) was set up by the Maritime Delimitation Treaty between The Republic of Colombia and Jamaica in November 1993. Colombia and Jamaica have territories on the same geological province, the Nicaraguan Rise. This area of shallow water stretches from Jamaica at its north eastern tip towards mainland Honduras and Nicaragua. Both countries will collaborate in understanding the geological systems which operate on the Nicaraguan Rise. The JRA is approximately 52,000 square kms. It will be the subject of geological and geophysical investigations prior to subdividing it into blocks for hydrocarbon exploration. The area will be jointly developed by Jamaica and Colombia with legal and jurisdictional arrangements overseeing operations.

While the search for oil and gas continues, Jamaica should increase its energy supplies by accelerating the evaluation, production, and marketing of indigenous alternative energy resources. Moreover, there is the need to strengthen energy management, from large industries to households. Energy efficiency and conservation, rational use of energy, and energy-induced technological changes in manufacturing industries are important in energy management. We must access alternative energy technologies as they become economic and available. Increasing energy efficiency is one of the most important requirements, and its merits should be widely promoted.

References

Meyerhoff, A.A., and Krieg, E.A. (1977) "Petroleum Potential of Jamaica". Special Report, Mines and Geology Division, Ministry of Mining and Natural Resources, p. 131.

Vision (1996) "Predictions Guarded in a 1996 Commodity Industry". *Petroleum Engineer International.* pp. 2–8.

JEBCO (2004) "The Realisation of the Hydrocarbon Potential of Jamaica Phase 1: Setting the Scene, Volume I/II."

4. | Coal

The World's Largest Fuel Source

COAL HAS BEEN USED AS a fuel for centuries all over the world and has a strong future. It ranks second to petroleum among all energy resources. World reserves of recoverable coal are approximately 850 billion tonnes, which may last another 150 years (World Energy Council, n.d.). The three most coal-rich countries in the world are the United States, Russia and China, as seen in Figure 4.1. Coal's availability, together with the relatively low costs required to generate power from coal, makes it a major substitute for oil in electric power generation. There is a need however to control emissions of the greenhouse gas, carbon dioxide, released by burning coal. In addition, about 50 per cent of the sulphur dioxide (So_2) and 30 per cent of the nitrogen oxides (NOx) released into the atmosphere result from the combustion of coal. In spite of the fact that coal production creates environmental hazards, coal will remain a major energy source for base load electricity in the future as oil and gas prices rise and 'clean coal technology' advances. Significant efforts are being made to limit the pollution caused by the use of coal as fuel by using scrubbers and other costly emission controls.

The conventional method of producing electricity from coal is to burn pulverized coal in a boiler and use the heat to vaporize water in steam tubes. The steam drives a turbine which converts energy into electricity by a generator. Typically, coal-fired plants have an efficiency of approximately 37 per cent. Emissions of waste gases (particularly So_2) are reduced by flue-gas scrubbing. When the gases come in contact with a limestone slurry or a similar absorbent, they react to form a compound that can be removed as solid waste. Scrubbers can reduce So_2 emissions

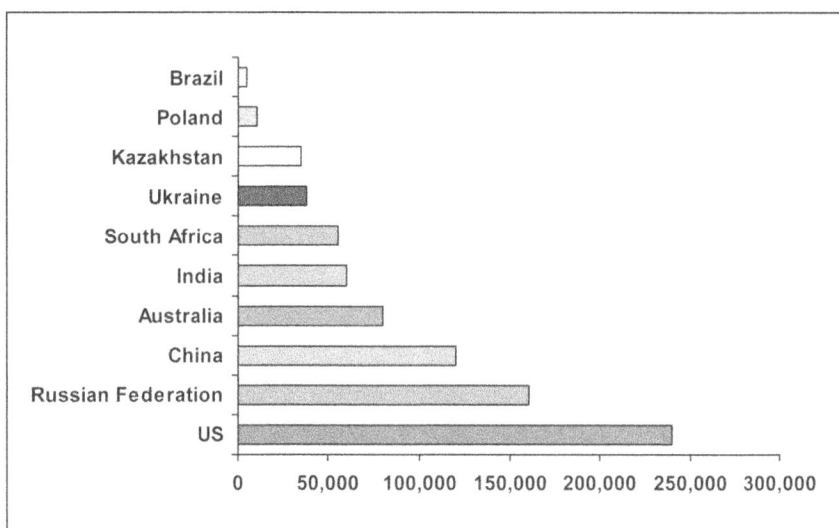

Figure 4.1   Countries with Major Coal Reserves
Source: BP Statistical Review of World Energy, 2008

by 50–85 per cent, with some loss of efficiency. A plant fitted with scrubbers runs at an efficiency of approximately 34 per cent. Only minor quantities of nitrogen oxide (NOx) gases are removed by scrubbers.

A new approach, the Integrated Gasification Combined Cycle (IGCC) system, is attractive in several ways (Figure 4.2). The major innovation is the transformation of coal into gas, a mixture of hydrogen and carbon with lesser amounts of hydrogen sulphide, and methane and carbon dioxide. Nearly all the hydrogen sulphide is removed by commercially available processes before the gas is burned. The synthesis gas then powers a combined cycle, the pressurized gases driving a turbine while vapour from the combustion chamber runs a conventional steam turbine. Efficiency claims for the IGCC system is about 42 per cent with electricity being generated at approximately 5 US cents per kWh, depending on location.

Repowering can improve efficiency and reduce unwanted emissions. Conventional steam boilers can be replaced by new types of combustion chambers such as the atmospheric fluidized-bed combustor, or the pressurized fluidized-bed combustor. Both systems are actively being used in the industry, and both burn coal with limestone or dolomite in a mixture suspended by jets of air. The mixture (fluidized bed) and the combustion gases envelop clusters of steam-generating tubes in and above the fluidized bed. In the pressurized system, boiler pressure is 6 to 16 times more

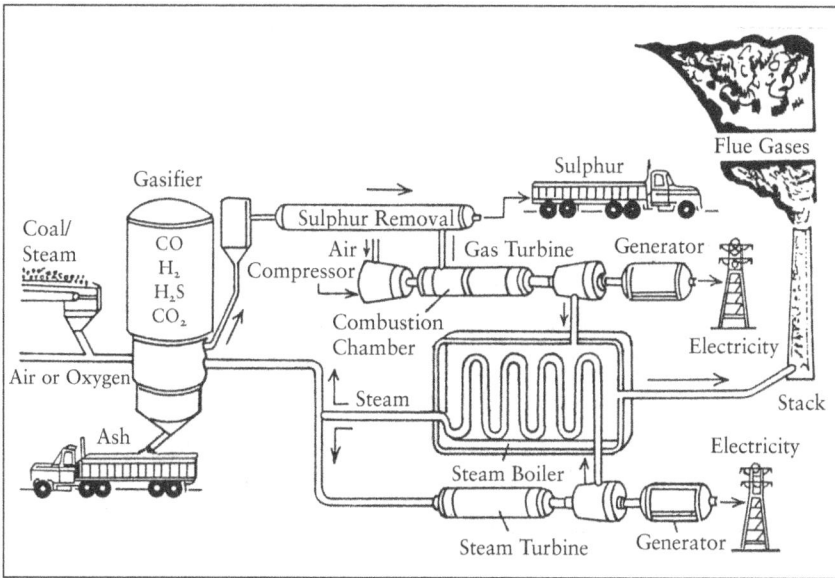

Figure 4.2 Integrated Gasification Combined Cycle (IGCC)
Source: Fulkeroon et al., 1990

than standard atmospheric pressure (Fulkerson et al., 1990). Greater efficiency is obtained by using the hot gases in the combustion chamber to run a turbine in addition to the standard steam turbine. Atmospheric fluidized-bed combustion is the least costly process with costs of approximately 5 US cents per kWh. Considerable commercial experience has accumulated with respect to atmospheric fluidized-bed combustion, using boilers in the size range suitable for Jamaica. All of these systems have costs low enough for the use of coal as a substitute for oil to be economically feasible, if the environmental risks are acceptable.

In terms of heating value, coal has consistently remained the least expensive source of fossil energy. Significantly, during the past 50 years the price of coal has not followed the same trend as coal production. From 1945 to 1969, the price of coal remained nearly constant at between US$4.00 and $4.60 per tonne. After adjustment for inflation, the value of a tonne of coal in 1969 had decreased to 60 per cent of the 1949 value. The coal industry was able to keep prices low and maintain the cost of energy from coal below that of petroleum by means of increased mechanization, increase in surface as against underground mining, and a trend towards fewer and larger mines. At the time of the 1973 oil price increase, the price structure of fossil fuels changed dramatically, moving ahead of inflation for the first time in several decades.

Coal, a fossil fuel is the largest source of energy for the generation of electricity worldwide as well as one of the largest global anthropogenic sources of carbon dioxide emissions

There are many factors, negative and positive, to be considered in coal use. Problems are yet to be resolved in mining, handling and shipping coal. The environmental considerations associated with combustion – acid emissions, sulphur and nitrogen oxides in the air we breathe, and carbon dioxide in the atmosphere – are all factors which adversely affect the future of coal utilization, and offset to some extent the economic benefits inherent in using coal.

National interests are best served by adequate and dependable supplies of the energy resources needed. The major increases in oil and coal prices in the recent past were the result of significant political events and not technological factors or normal market reaction. Of the many disparate factors that influence the world energy markets, the most difficult to predict are the national paroxysms within the major oil-producing areas. The resulting instability of the international energy economy has had a greater impact on coal use than any technological or environmental factor. Supply uncertainties in oil, as well as lower cost and ready availability, are causing a swing in the direction of increased coal utilization. But as oil prices increase, there is a tendency for coal prices to follow suit.

Predicting the future use of coal worldwide is complicated by the wide discrepancy in per capita energy use between developed and developing countries. Energy consumption in the highly industrialized countries has leveled off, mainly due to the use of energy-efficient technologies in those countries. However, the energy consumption of developing countries rises with their economic growth in contrast to energy consumption in the USA, for example, where it has remained fairly level for the past 15 to 18 years. The greatest growth will be in China where increased coal use is expected to be significant over the period 2010–2020. China has the third largest coal reserves in the world and continues record-breaking annual production in excess of one billion short tons of coal.

Environmental considerations will probably be the most restraining factor on coal use in the future and so the introduction and use of clean coal technologies will be of special significance. These technologies will make the burning of coal for electricity more efficient while reducing atmospheric pollutants.

Coal presently provides 23 per cent of primary energy and 39 per cent of electricity generation globally. Further, there are reserves of coal that will last upward of 100 years. For this reason, coal will be an important part of the global energy future. There are a number of techniques to make the use of coal for energy more environmentally friendly.

- Coal cleaning by 'washing' has been standard practice in developed countries for some time. It reduces emissions of ash and sulfur dioxide when the coal is burned.
- Electrostatic precipitators and fabric filters can remove 99 per cent of the fly ash from the flue gases – these technologies are in widespread use.
- Flue gas desulfurization reduces the output of sulfur dioxide to the atmosphere by up to 97 per cent, the task depending on the level of sulfur in the coal and the extent of the reduction. It is widely used where needed in developed countries.
- Low-NOx burners allow coal-fired plants to reduce nitrogen oxide emissions by up to 40 per cent. Coupled with re-burning techniques, NOx can be reduced by 70 per cent and selective catalytic reduction can clean up 90 per cent of NOx emissions.
- Increased efficiency of plant operations – up to 45 per cent thermal efficiency – now means that newer plants create less emissions per kWh than older ones.
- Advanced technologies such as Integrated Gasification Combined Cycle (IGCC) and Pressurized Fluidized Bed Combustion (PFBC) will enable higher thermal efficiencies still – up to 50 per cent in the future.
- Ultra-clean coal from new processing technologies that reduce ash below 0.25 per cent and sulfur to very low levels, mean that pulver-

ized coal could be fed directly into gas turbines with combined cycle and burned at high thermal efficiency.

• Gasification, including underground gasification in situ, uses steam and oxygen to turn the coal into carbon monoxide and hydrogen.

A 300 MW coal project in Jamaica would involve an infrastructural cost for receiving and managing the product, which is approximately US$150 million. Jamaica has no high quality coal reserves. There are non-commercial seams of lignite associated with the tertiary yellow limestone strata surrounding parts of the cretaceous central inlier. Sources of coal for Jamaica include the United States and Columbia, the latter nation being the nearest source of high grade coal for Jamaica. Colombia has approximately 7,500 million short tonnes (Mmst) of recoverable coal reserves consisting of high-quality bituminous coal and a small amount of metallurgical coal. The country has the second-largest coal reserves in South America, behind Brazil, and most of those reserves are concentrated in the Guajira peninsula in the northeast and the Andean foothills. Colombia's coal is relatively clean-burning, with a sulfur content of less than 1 per cent. It is likely that Colombia's coal production will continue to increase in coming years, as exploration and profitable developments continue throughout the north and interior of the country.

The exploration of coal will continue in many countries around the world, and coal will maintain a dominant position in the energy future for the world for some time. Coal is growing in many countries, including population-dense and fast growing countries such as China and India. The pace of future coal production will depend on the speed with which innovation reduces the cost of using alternative energies. Though coal has clearly made progress in developing technologies that mitigate its effects on the environment, it is likely that debates will continue on the likelihood of there ever being a "green" coal.

References

Fulkerson, W., Judkins, R.R., and Sanghui, M.J. (1990) "Energy from Fossil Fuels". *Scientific American* 263 (3) pp. 129–135.

Spence, D.B. (2005) "Coal-Fired Power in a Restructured Electricity Market". *Duke Environmental Law & Policy Forum* 15 (2) pp. 187–195.

World Energy Council (n.d.) "Survey of Energy Resources 2007". Retrieved from: http://www.worldenergy.org/publications/survey_of_energy_resources_2007/coal/627.asp.

5. | Natural Gas

The Cleanest Fossil Fuel

NATURAL GAS HAS MANY RESIDENTIAL, commercial and industrial uses. Like oil and coal, it is a fossil fuel, meaning that it is essentially the remains of plants, animals and microorganisms that lived millions of years ago. Natural gas is regarded as a transition fuel as we move away from oil towards a new energy agenda that will include renewables.

Natural gas is found in reservoirs underneath the earth and is commonly associated with oil deposits. Consequently, major oil-producing countries are also major producers of natural gas. Figure 5.1. shows the nations with the largest natural gas reserves.

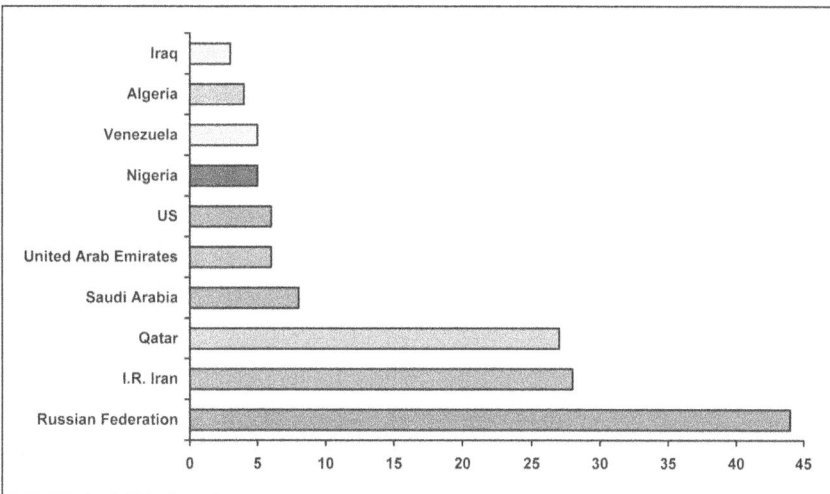

Figure 5.1   Countries with Major Natural Gas Reserves
Source: BP Statistical Review of World Energy, 2008

Natural gas is a combustible mixture of hydrocarbon gases and in its purest form is almost all methane. Methane is a molecule made up of one carbon atom and four hydrogen atoms, and is referred to as CH_4. Natural gas can also include various amounts of ethane, propane, butane and pentane. Table 5.1 outlines the typical makeup of natural gas before it is refined.

Table 5.1   Typical Composition of Natural Gas

Methane	CH_4	70–90%
Ethane	C_2H_6	0–20%
Propane	C_3H_8	0–20%
Butane	C_4H_{10}	0–20%
Carbon Dioxide	CO_2	0–8%
Oxygen	O_2	0–0.2%
Nitrogen	N_2	0–5%
Hydrogen sulphide	H_2S	0–5%
Rare gases	A, He, Ne, Xe	Trace

Source: Natural Gas.org

Liquefied Natural Gas and Compressed Natural Gas

Natural Gas is utilized commercially as Liquefied Natural Gas (LNG) or Compressed Natural Gas (CNG). Liquefied Natural Gas is a naturally occurring mixture of hydrocarbons, mainly methane or CH_4 that have been purified and condensed to liquefied form by cooling cryogenically to -162 degrees Celsius. At atmospheric pressure, LNG occupies only 1.610 the volume of natural gas in vapour form. Methane is the simplest molecule of the fossil fuels and can be burned very cleanly. It has an octane rating of 130 (compared to gasoline at 90–95 octane) and contains excellent properties for spark-ignited internal combustion engines and for use in driving turbines, heating boilers and furnaces. LNG must be kept at very cold temperatures, and is stored in double-wall vacuum-insulated pressure vessels. Compared to conventional fuels, LNG flammability is limited. It is non-toxic, odourless, non-corrosive, and non-carcinogenic. It presents no environmental threats to soil, surface water or groundwater.

LNG is used primarily for international trade in natural gas and for meeting seasonal demands for natural gas. It is produced mainly at LNG storage locations operated by natural gas suppliers and at cryogenic (liquefaction) extraction plants in gas-producing states. The countries with the largest LNG consumption are Japan and South Korea; together they import nearly 60 per cent of the LNG exported in the world market. Despite the success of individual LNG projects and the regional importance of LNG, overall LNG accounts for only 6.5 per cent of the world natural gas consumption and has a minor influence on world patterns of gas consumption. This influence is growing rapidly as natural gas producers seek to curtail flaring gas and to monetize gas resources.[1]

The environmental benefits of LNG include significant reductions in carbon dioxide and particulate commissions when compared with other fossil fuels. In addition, LNG is an excellent fuel for producing low cost power. Its use will promote industries that can derive benefit from cheaper power in the vicinity of the receiving terminal. The importation of LNG into Jamaica could stimulate new petrochemical and allied industries derived from methane, and these industries could be sited in an industrial park adjacent to the import terminal.

The other commercial form of natural gas, CNG, is a fossil fuel substitute for gasoline (petrol), diesel, or propane fuel.[2] CNG is a more environmentally clean alternative to these fuels and is much safer than other fuels in the event of a spill.[3] CNG is made by compressing natural gas (which is mainly composed of methane [CH_4]), to less than 1 per cent of its volume at standard atmospheric pressure. It is stored and distributed in hard containers, at a normal pressure of 200–220 bar (2,900–3,200 psi), usually in cylindrical or spherical shapes.

CNG is used in traditional gasoline internal combustion engine cars that have been converted into bi-fuel vehicles (gasoline/CNG). Natural gas vehicles are increasingly used in Europe, South America and Asia due to rising gasoline prices. In response to high fuel prices and environmental concerns, CNG is beginning to be used in light-duty passenger vehicles and pickup trucks, medium-duty delivery trucks, transit and school buses, and trains.

Methanol Fuel

Another product of natural gas, methanol,[4] is being considered as a fuel and is now being regarded as a possible alternative to LNG. Like LNG, methanol is manufactured from natural gas with a slightly higher capital cost per unit of energy than LNG. It is however cheaper to transport than LNG. Methanol has lower energy content than LNG, equivalent to approximately 66 per cent of the gas consumed in its production. On the positive side, methanol is a clean fuel suitable for gas turbines, gasoline engines and in new fuel cell technologies. Methanol is competitive, creating opportunities for its manufacture in natural gas-rich regions. It has the potential to become an important fuel over the longer term as gas-producing counties seek to monetize all available gas supplies. Methanol will also have a market in PetroChemicals, pharmaceuticals, as well as emerging fuel cell technologies.

Hydrates

Gas Hydrates are nonstoichiometric crystalline compounds that belong to the inclusion group known as clathrates. Hydrates occur when water molecules attach themselves together through hydrogen bonding and form cavities which can be occupied by a single gas or volatile liquid molecule. The presence of a gas or volatile liquid inside the water network thermodynamically stabilizes the structure through physical bonding via weak van der Waals forces.

Naturally occurring hydrates are being looked upon as a future energy source and a potential global climate hazard. These naturally occurring hydrates, containing mostly methane, exist in vast quantities within and below the permafrost zone and in sub sea sediments. At present the amount of organic carbon entrapped in hydrate exceeds all other reserves, including fossil fuels, soil, peat, and living organisms.[5]

Carbon dioxide hydrate is another important hydrate. Carbon dioxide, like methane, is a component of natural gas and may form hydrates in oil reservoirs during enhanced oil recovery, thereby causing complications. Carbon dioxide is also a major component in the emissions of thermal power plants which contribute to global warming. Work is now being conducted on capturing carbon dioxide from thermal power plants, and subsequently transforming carbon dioxide into hydrates in order to sequester them in the deep ocean, ultimately to improve atmospheric conditions.

Natural Gas Prices

Although natural gas prices have taken a general upward trend in tandem with crude oil prices over the past five years, they also have shown considerable independent movement. Brown and Yuckel (2008) find a direct relationship between natural gas and crude oil pricing based on the relationship between the two sources of energy.

Natural gas prices, like other commodity prices, are driven by supply and demand fundamentals. The supply for Natural Gas is mainly driven by the following factors:

- Weather conditions
- Storage
- Natural Phenomena
- Imports
- Pipeline Capacity
- Gas Drilling Rates
- Technical Issues

The demand for Natural Gas is mainly driven by the following factors:

- Weather conditions
- Economic Growth
- Storage
- Demographics
- Fuel Competition
- Exports

The reader will note that weather affects both supply and demand. The future of the supply of natural gas is largely dependent on the weather in many regions of the world. For example, the natural gas operations in The Gulf of Mexico are vulnerable to damage due to the increasing number of hurricanes in that region. Immediately after Hurricane Katrina in 2005, natural gas prices became high because of the general destruction.

Even though natural gas is one of the more volatile commodities markets, it remains an excellent option because of its status as a relatively clean form of energy. Research on natural gas is vibrant. Hydrate research especially, continues apace and is recognized as an important field not only because of the hazards posed by gas hydrates, but also because of great possibilities that hydrates provide for humankind.

References

Brown S., and Yuckel K. (2008) "What Drives Natural Gas Prices?" *The Energy Journal* 29 (2) pp. 45–59.

Naturalgas.org "Background". Retrieved from: http://www.naturalgas.org/overview/background.asp.

6. | Nuclear Power

Safety and Security are Controversial Concerns

NUCLEAR ENERGY IS RELEASED BY the splitting (fission) or merging together (fusion) of the nuclei of uranium atoms. Nuclear energy was first discovered by French physicist Henri Becquerel in 1896, when he found that photographic plates stored in the dark near uranium were blackened like X-ray plates.[1]

Of all the major new sources of energy, nuclear fission has received the most financial support, and as a result, its technology is the best developed. Nuclear energy presently accounts for approximately 15 per cent of the world's electricity; over 450 nuclear reactors are connected to the global electricity grid in 30 countries. There is currently a renewal of interest in nuclear energy, particularly in the countries of the former Soviet Union and the European Union. France produces the highest percentage of its energy from nuclear reactors, almost 80 per cent.

Nuclear reactors in use today have a minimum efficient size of approximately 600 MW, which is not far below the daily power demand from the utility company in many small nations. A switch to nuclear power in small nations such as Jamaica would cause near total dependency on one power supply source, strategically a sensitive situation. Nonetheless, as Jamaica's energy demand increases and as smaller nuclear plants become viable, nuclear power could become a part of the energy mix.

Nuclear fission has substantial advantages over fossil fuels. Nuclear power plants do not emit particulates, sulphur oxides, or other combustible products. Nuclear fuel (uranium) is a more compact source of energy than coal and requires less mining activity. The water pollution and land disruption associated with mining are correspondingly reduced.

Table 6.1. Nuclear Electricity Generation and Capacity in Selected Countries as at End 2008

Country	Number of nuclear units connected to the grid	Nuclear electricity generation (net TWh) 2007	Nuclear percentage of total electricity supply
Belgium	7	45.9	54.1
Canada	20	88.6	14.7
Czech Republic	6	26.2	32.2
Finland	4	22.5	29.0
France	59	418.6	76.8
Germany	17	133.2	23.2
Hungary	4	13.8	37.2
Japan	55	251.6	25.6
Korea	20	136.3	35.2
Mexico	2	10.4	4.4
Netherlands	1	4.0	4.0
Slovak Republic	5	14.1	54.9
Spain	8	53.4	17.8
Sweden	10	64.3	47.4
Switzerland	5	26.3	39.9
United Kingdom	19	57.3	15.7
United States	104	806.0	19.4
Total	346	2 172.5	21.6
OECD America	126	905.0	18.1
OECD Europe	145	879.6	25.8
OECD Pacific	75	387.9	23.3

Source: Nuclear Energy Data 2009

Transportation costs for nuclear fuels are lower. Fission can also help to replace dwindling fossil fuels as a source of energy and so conserve them for use as chemicals. The greenhouse effect is a strong argument for increasing the use of nuclear power, particularly if the safety and waste problems can be overcome.

Risk and Risk Perception

Despite the advantages of nuclear energy, there are significant concerns about its use, as listed below.

- There is always the remote possibility of a serious accident.
- Dangerous human actions ranging from carelessness to sabotage are possible with nuclear power systems.
- The consequences to human health and the environment of any major release of radioactive substances make nuclear fission potentially the most hazardous of all energy sources.
- Safeguarding the fissionable materials used as reactor fuels presents difficulties.
- Unresolved problems still exist in the long-term storage and ultimate disposal of radioactive wastes.

Clearly, the risks associated with nuclear energy need to be studied carefully before investing in this form of energy.[2] The public perception of the risk is arguably the greatest problem. According to Walker (2000), researchers have offered several explanations for the high levels of public

Nuclear power is growing internationally primarily because of minimal polluting emissions, the only drawback being the safe disposal of the nuclear waste

fear of radiation and the concomitant fear of nuclear power. A prevalent theory is that radiation is terrifying because it is undetectable to the human senses. Other researchers have cited additional considerations that fuel public fear, one being the involuntary imposition of risk from the radiation released by nuclear power facilities. Individuals view the technology as something beyond their control: powerful external forces such as the nuclear industry, utilities and governments determine the level of risk, weigh the hazards against the benefits, and set standards of safety.

Though the risk may be proportionately lower than statistically more dangerous activities that individuals assume voluntarily, such as driving, risk perception is also affected by what is seen as more dramatic. To make matters more alarming, a radiological accident has no definite end because uncertainty about the long-term effects of the exposure continues. The lingering spectre of a catastrophic radiation accident such as Chernobyl on April 26, 1986 has dominated public attitudes toward the safety of the technology. In some countries, there are regulatory and licensing uncertainties which may also be related to risk perception. In Sweden, a referendum called for the cessation of nuclear power generation by 2010. In Germany and Switzerland, there has been a moratorium on new plants.

Economic Issues

Despite the risk and risk perception, the rising costs of energy supplies have reframed the debate in terms more favourable for nuclear energy advocates. Consequently, the discussion on whether governments should encourage the construction of new nuclear power plants is being reintroduced in Australia, the United States and other countries. What occurs in China is of particular interest because of its large population. To achieve its goal of generating 4 per cent of electricity from nuclear power, China will add more than 30 new nuclear plants by 2020 to the 11 currently in operation or under construction. This in itself indicates that nuclear energy may not be on a rapid slide downward as many had predicted.

The optimism about the economic benefits of nuclear power in the late 1970s was based on the rapid increases in petroleum prices at that time. Those increases were not sustained after the mid-1980s. The operating costs of coal-fired plants decreased, bringing them closer to the projected costs for nuclear power. Some governments and utility organ-

izations began to use higher discount rates in evaluating investments in nuclear power because of its perceived uncertainty and risk, thus increasing the computed costs for such capital-intensive technologies.

The main hope of nuclear power in the future lies in reducing costs by simplifying and standardizing plant design. Through design standardization, development of compatible safety and regulatory standards, and the merger of international nuclear manufacturers, the industry is becoming more consistent and globalized. The effort is motivated in large measure by the shrinking markets for new nuclear plants.

A new generation of fission energy is predicated on the commercial arrival of small pebble-bed nuclear reactors. These are helium-cooled reactors that contain an array of small tennis ball-sized 'pebbles'; with just 9 grams of uranium per pebble, which creates a low power density reactor. The low power density and large graphite core provide inherent safety features. The peak temperature reached even under the complete loss of coolant in the core cooling system is significantly below the temperature that the fuel melts, so these reactors are inherently safe. Pebble bed reactors are expected to become available at some point between 2020–2025. They would be economically viable in the size range of 90–200 MW and would be suitable as a least cost economic option for many developing countries, including Jamaica.

References

Leuwen, J., and Smith, P. (2008) "Nuclear Power: the Energy Balance", Chaam, Netherlands.

OECD Nuclear Energy Agency (2009) "Nuclear Energy Data 2009/Données sur l'énergie nucléaire 2009", Published: 04-SEP-09, 120 pages.

Sangster, A.W. (1984) *Energy and Our World* Kingston: College of Arts, Science and Technology, Kingston.

Walker, J.S. (2000) *Permissible Dose: A History of Radiation Protection in the Twentieth Century.* California: University of California Press.

World Nuclear Association (n.d.) "Nuclear Power in the World Today". Retrieved from: http://world-nuclear.org/info/info1.html

7. | Renewable Energy

Fact or Fiction

The Stone Age did not end because of a lack of stones, and the Oil Age will not end because of the lack of oil.
– Sheikh Zaki Yamani, 1993

SUPPLIES OF OIL AND GAS were subject to uncertainty and price fluctuations during the 1970s and 1980s. With the 1990s came a respite from the wild swings in prices and availability but the lessons from previous decades indicated that long-range plans should not be made on the hopes of continuing stability in the energy markets. The concern over stable prices and supplies and the awareness of environmental problems caused by the use of conventional energy resources have made alternative energy resources into options to be seriously considered. Energy technologies should be selected based on societal benefits and costs, because there are more risks and greater impacts on society than just the cost per kilowatt hour charged by the utility company.

Many developing countries, including Jamaica, are endowed with an abundance of sunshine and trade winds. These are free, renewable and inexhaustible sources of energy which can be tapped to meet a part of the country's energy needs. Jamaica is almost totally dependent on imported oil for its energy requirements. The cost of electricity to the domestic consumer in Jamaica is high in comparison to other Caribbean countries such as Trinidad and Tobago and neighbouring countries in Central and South America. The reliability of the electrical power supply has improved but the fluctuations in voltage are still excessive and lengthy. The Jamaican utility company is dependent in large measure on

Table: 7.1   Summary of Some Renewable Energy Programme Goals

Factor	Electricity Prices (US$ Per Kilowatt Hour)	
	Status 2009	Mid Term Goals (2015–2020)
Photovoltaics	18–25 cents	8–14 cents
Solar Thermal	10–25 cents	7–15 cents
Wind Energy	6–18 cents	4–12 cents
Biomass	6–22 cents	5–12 cents
Average cost to Jamaican utility to generate electricity*	8.6 cents	7.5 cents

*Author's estimate

old power generation equipment which has to be replaced in the near future.

Can renewable energy make economic sense in the immediate future? The answer is that some technologies are presently economic whereas others require technological improvements in order to become more economic. The cost effectiveness of various renewable energy technologies will be discussed further in the following chapters.

Table 7.1 shows the projected prices of some renewable energy programmes in Jamaica.

Wind Energy

Wind is solar power that has already been converted into mechanical power, therefore conversion to electricity can be accomplished efficiently. Wind turbines have been used for centuries to pump water and grind grain and sugar cane. During the past 30 years, the wind turbine market has evolved from small machines used for domestic and agricultural purposes to interconnected wind farms. A wind farm is a group of turbines sited at a single location, serving a common application to generate electricity. A wind speed averaging at least 21 km/h (13 mph) is required.

The power produced by wind turbines is attractive to the utility company because periods of high wind (between 10:00 a.m. and 3:00 p.m.)

coincide with high air conditioning loads. Since peak power is the most costly to produce, wind energy is particularly valuable because it is available at the time of peak demand.

Biomass

Biomass such as sugar cane, sorghum and leucaena (a fast-growing tree) are excellent converters of solar energy to usable energy. Biomass for energy will not play a significant role in oil substitution in the short term in most developing countries. The assessment, exploitation and development of biomass resources should co-exist with the optimization of oil use and energy conservation. The burning of trees for charcoal production is a biomass energy source that has to be strictly controlled because of the negative effects of deforestation.

Direct Solar Systems

Solar water heating on a domestic, hospital, and commercial scale has been demonstrated in many countries. Solar thermal electric systems have long been known. The sun is concentrated by mirrors onto a receiver that contains a fluid. The heated fluid is then used to produce steam which drives a turbine generator. The fluid could also be a gas which operates an engine directly. Costs are still high when compared to conventional power generation, between 10 and 20 US cents per kilowatt hour, which is nearly twice the cost of fossil fuel plants. The technology is modular, suited to mass production, although economies of scale have yet to be achieved. Solar thermal and photovoltaics show great promise for the future. Photovoltaic energy is produced from a normal silicon-based solar cell which generates electricity from sunlight. As a rule of thumb, one square metre of solar cell will produce between 50 and 150 Watts of electricity, depending on the sun's radiation in a given area.

Why Renewables?

The argument for alternatives to fossil fuels, such as the nuclear power schemes of 1955 to 1975, was that a backstop technology was required in case we ran out of fossil fuels. But the world supply of fossil fuels has proven to be larger than previously thought. Even allowing for growth of demand in developing countries, there is enough coal and oil to last at least another hundred years.

Renewable energy is just beginning to compete with fossil and nuclear fuels, especially in small-scale applications. Investment in renewable energy is an integral part of any precautionary policy. Energy efficiency alone is not an answer. For instance, a 25 per cent level of savings from energy efficiency would in 40 years still leave the world's energy consumption equal to approximately three times its present level.

Solar energy has no net emissions of carbon dioxide, particular matter, sulphur dioxide or nitrous oxide. Nonetheless, there is a perception that solar energy is too diffuse to harness effectively and requires too much land space. Although solar-thermal and photovoltaic projects have larger land requirements than coal or oil-fired plants at the point of electricity generation (that is, excluding the mining area), they use considerably less land than hydropower schemes. There is also much flexibility in the choice of site. The plants are modular and can be located in sparsely populated areas without having to compete with agriculture, forests or people for land. The yield of energy from photovoltaic sources, for example, assuming 10 per cent conversion efficiencies, is about one hundred times that of biomass energy forests.

The developments in photovoltaics in the past twenty years have been rapid. There have been immense technical achievements, considering that public policies have actually worked against renewable energy by heavily subsidizing fossil fuels. There has been significant market growth for remote and off-grid applications including village lighting, health clinics, telecommunications, security lighting and water pumping. Over 60 large manufacturers are investing in the technology, many preparing for a second stage of larger-scale production in the expectation of a larger market and further cost reductions. One problem with photovoltaic electricity is the high cost of storage, especially in off-grid applications. The lead acid battery is still the equipment most used but it is impractical for large-scale operations. Research on fuel-cell to be used in electric cars has spin-offs in its potential application to the storage of photovoltaic energy.

Renewable energy will benefit from the adoption of strictly commercial policies for energy. In addition, it will welcome the termination of government controls and subsidies on energy pricing, and the end of public utility monopolies in electricity generation so that small independent producers can generate and sell electricity to the grid. A projection of the worldwide contribution of renewable energy is illustrated in Figure 7.1.

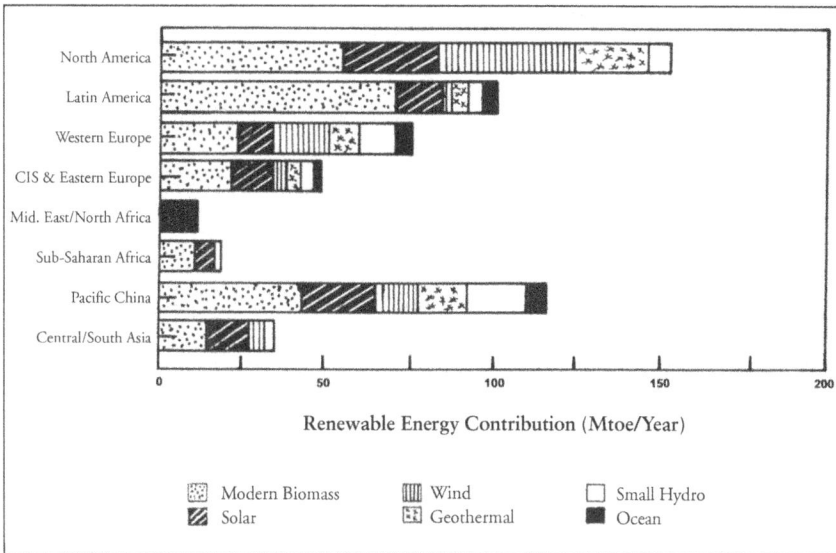

Figure 7.1 Renewables Distribution in 2020 – Current Policies Scenario
Source: World Energy Council

Policies and Strategies to Drive Renewable Energy

The following policies and strategies are required to drive renewable energy:

Regulations
- Renewable Performance Standards
- Performance standards for new facilities, e.g. GHG emissions no greater than combined cycle gas

Green Power Purchasing Requirements
- Interconnection standards
- Net metering rules
- Generation disclosure rules
- Contractor licensing
- Equipment certification
- Renewable access laws/guidelines/zoning codes/building permit requirements

Taxes and Charges
- Corporate tax credits/depreciation rules
- Personal income tax credits

- Sales taxes exemptions
- Property tax credits/charges

Fiscal Incentives

- Feed in tariffs
- Rebates
- Grant programmes
- Loan programmes
- Bonds
- Production incentives
- Government purchasing programmes
- Equity investments, including venture capital
- Insurance programmes
- Carbon taxes

Other Economic Instruments

- Renewable energy certificates (tradable)
- CO_2 Emission trading programmes

Research and Development

- Tax credits
- Grants
- Public/Private partnership
- Private investment
- Domestic Bank Financing

Policies relating to renewable energy are usually coupled with policies relating to education, industrial development, transportation, environment, waste management, land use and the goal of sustainable development.

It is clear that renewable energy will become a part of development plans of both developed and developing countries. Its role will be especially significant in oil deficient economies. Market applications of renewable energy become more diverse as the public becomes more familiar with these technologies which will become, with reducing costs, justifiable on economic criteria alone. We will then see an example of an approach to solving environmental issues that creates a welcome positive economic surprise. That will be fact, not fiction.

8. | Wind Energy

The Fastest Growing Renewable Technology

WIND ENERGY IS A RENEWABLE energy resource that has come of age during the last two decades. If wind energy fulfils its promise, it could provide as much as five per cent of the world's electricity by the year 2035. Wind energy provides a better energy balance, an indigenous source of supply, more local investment and employment, and less reliance on imports than oil and gas. Wind energy is becoming more attractive worldwide as the environmental and other external costs from both fossil fuels and nuclear sources are becoming higher. Wind has advantages in non-fuel use, capacity savings, emission savings and lower operations and maintenance costs.

Wind energy has developed rapidly in terms of cost-effectiveness and technical performance. Wind energy has the advantage of being abundant and sustainable. No heat, air or water pollution is produced when wind power is converted to energy. In contrast to thermal plants (coal, nuclear, fuel oil), it does not require water for the production of electricity. Its disadvantages are the low density of the energy and the variability of the wind. Wind is free, but the capture of that energy is not free. The low density means that the initial cost of wind turbines is relatively high. If on-site storage is needed, then the initial cost can double. The progress in the development of wind energy over the last ten years gives assurance that a gradual shift to alternative energy will be feasible.

History

Wind has been harnessed for power for a long time. As early as 200 BC, the first wind turbines were being used to grind grain in Persia. These were simple vertical-axis machines. Later, horizontal-axis windmills rigged with jib sails were developed, and some are still used today in the Mediterranean area. The Dutch took the lead in improving windmill design, and by the fourteenth century they were using windmills to drain the marshes and lakes of the Rhine River delta. In the middle of the twentieth century, small windmills became an energy source for water pumping and electric power generation. Between 1930 and 1950 a few large, experimental, wind powered electric systems were built in Europe and North America. By the second half of the 1950s, electric wind generators were displaced by inexpensive, centralized, oil and coal electric power generation. When a 300 per cent increase in world oil prices occurred in 1973–74, wind turbines were again envisioned as an alternative source of energy. Since that time, research and development of wind energy by governments and industries has resulted in the emergence of the modern international wind energy industry.[1]

The Caribbean has a long and interesting history with respect to windmills. To illustrate, windmills were introduced in Barbados in the seventeenth century for the purpose of crushing sugar cane. By 1750, 356 windmills had been installed and by 1846 the number had risen to 506. After a long period of decline, only one windmill remained operational in Barbados in 1946 (Buisseret, 1975). In Jamaica, water power was more widespread than wind power; however, by 1804 there were 86 windmills notably in Hanover, St James, Trelawny, St Mary, St Thomas and southern Clarendon (Buisseret, op.cit.), used almost entirely used to grind sugar cane. Several windmills were put into operation during the 1930s to early 1960s to pump water from shallow wells in the coastal plains of Clarendon and St Elizabeth. Poor maintenance and reliability caused these pumps to be displaced by electric and diesel pumps in the 1960s and 1970s. Probably fewer than five windpumps now operate in Jamaica. The first electric wind turbine in Jamaica, a small unit generating 225 k\W, was installed at Munro College in St Elizabeth, and became operational on December 20, 1995 (Figure 8.1). This unit became inoperable after four years, but there are plans to add new wind turbines at the Munroe site.

The Economics of Wind Energy

While traditional energy costs have been rising, wind energy costs have been declining. Larger and more efficient manufacturing, advances in technology and improved experience with wind turbines have contributed to this trend.

Wind technology is already the least costly source of renewable energy, less costly than of photovoltaics and solar thermal energy, and has become the most economic new source of electricity. When located at appropriate sites, the costs of wind energy are now competitive with the full cost of power from conventional sources such as coal and oil. Wind energy provides other economic and social benefits such as a greatly reduced environmental impact, reduced dependence on fossil fuels, and flexibility in electric-power planning as a result of the simplicity and modularity of wind turbines. These factors allow more options to be considered in layout and expansion on suitable sites. Another socioeconomic benefit is the compatibility of wind energy with other land uses such as animal rearing and agriculture. Policy makers are realizing that these benefits have value and should be considered when calculating energy costs.

With improved technology, the cost of energy from properly maintained small and intermediate sized wind turbines is expected to drop to less than 5 US cents/kWh in 2009 at sites with optimal wind speeds. Yet a wind turbine is economically feasible only if its overall earnings exceed its overall costs within a given time period. The time taken to reach the

Wind energy is the fastest growing renewable resource with an incremental growth of about 17 per cent per year

point at which earnings equal costs is called the payback period. The relatively large initial cost of installation means that this period could be over a number of years. A short payback is of course preferable and a payback of five to seven years is acceptable. Longer payback periods should be viewed with caution by investors. Investors should also note that between 65 and 70 per cent of the capital cost is dedicated to the cost of the turbine. The most favourable sites would have average wind speeds of 8 metres/second, installed capital costs of US$1500/kW and a rate of return on investment of 20 per cent.

Operating Principles

Winds are generated by changes in atmospheric pressure induced by changes in earth and atmospheric temperature. They are further affected by the earth's rotation and by frictional encounters with topographic features such as mountain ranges, and are therefore not uniformly distributed. Wind resources are classified according to wind-power density, ranging from class 1 (the lowest) to class 7 (the highest). Good wind resources of class 4 and above, with an average wind speed of at least 21 kilometres per hour (13 mph), are obtained in many areas of Jamaica and other Caribbean islands.

Wind turbines capture energy from the wind with a rotor which usually consists of two or three blades mounted on a shaft. The spinning blades rotate a generator to produce electricity. The turbines are mounted on towers to optimize the capture of the wind's energy because wind speed is generally slower and the wind more turbulent closer to the ground.

Although various configurations and designs exist, wind turbines are generally of two types: vertical axis and horizontal axis. Vertical-axis wind turbines resemble an upright eggbeater, in which the axis of rotation is roughly perpendicular to the wind stream. Horizontal-axis wind turbines resemble a traditional windmill, in which the axis of rotation is nearly parallel to the wind stream (Figure 8.1). A blade or rotor converts the wind to rotational shaft energy which produces electricity through a gearbox and generator (Figure 8.2). The blades can have variable or fixed pitch, twist or other airfoils. For units connected to the utility grid the generators can be synchronous, induction, variable frequency alternator or direct current with an inverter. Although wind turbines can be sited

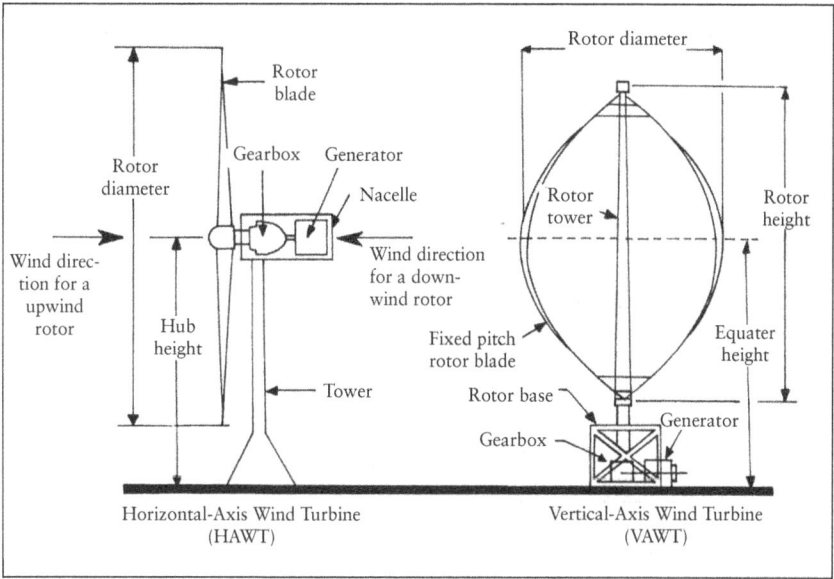

Figure 8.1 Schematic Overview of Wind Turbine Components
Source: Goodbody and Thomas-Hope, 2002

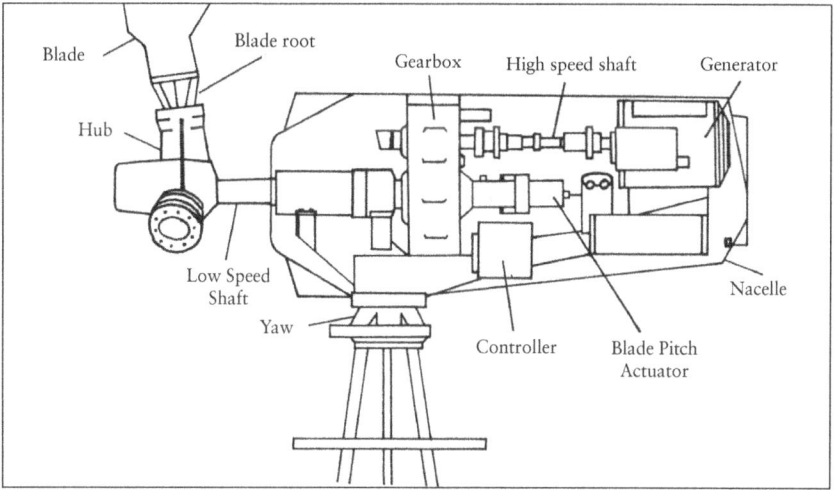

Figure 8.2 Components of a Wind Turbine

individually or as stand-alone systems, there are operating advantages to siting the turbines in an array to form a wind farm connected to the power grid.

As Figure 8.2 shows, the yaw system controls the directional orientation of the rotor, keeping the turbine facing into the wind. The yaw system usually contains a bearing surface for directional rotation of the turbine, a drive system to move the turbine when the wind changes direction, an error-sensing system, and a mechanical brake used during system servicing. The yaw drive consists of a hydraulic motor, planetary gear and hydraulic damping. The yaw drive, gearbox, control system and generator are housed in a nacelle (usually made of fibreglass). This protects the turbine from environmental exposure including rainfall, airborne particles, and sunlight.

The electrical system of a wind turbine is usually composed of a low voltage subsystem which works at 460–480 V. This is attached from the downtower facilities of individual turbines to step-up transformers at the windplant substations where the power is increased to 220–240 kV for transmission to the grid. The output of a wind turbine varies with the wind speed through the rotor. At low wind speeds, less than 16 kilometres per hour (10 mph), the power output drops off sharply. Thus wind-resource evaluation is a critical element in estimating turbine performance at a given site.

The energy available in a wind stream is proportional to the cube of its speed, so that doubling the wind speed increases the available energy by a factor of eight. Also, the wind resource is not often a steady consistent flow. It varies with the time of day, season, type of terrain and height above ground. Wind speed apart, the most important parameter in determining energy production is the rotor area. Energy increases in proportion to the square of the radius. A larger generator does not necessarily mean more energy production because efficiency at low wind speeds will change with generator size.

Usually the energy output of wind turbine systems are given per year for comparison. Mean energy output per square metre swept rotor area over a year can be expressed by the formula:

$$E = b\,(V3),\ (\mathrm{kWh/m^2\ per\ year})$$

where b is the performance factor and V the annual mean wind speed. The factor b measures the performance quality of a wind turbine. However, b is directly dependent on the annual mean wind speed and wind-speed distribution. Hence b is a variable throughout the world. Because the energy in the wind increases as the cube of the wind speed, all wind turbines must be able to release (dump) power at high wind speeds. The ways of doing so include:

- changing the aerodynamic efficiency of the blades (rotor) by variable pitch, spoilers, or operation at constant rpm

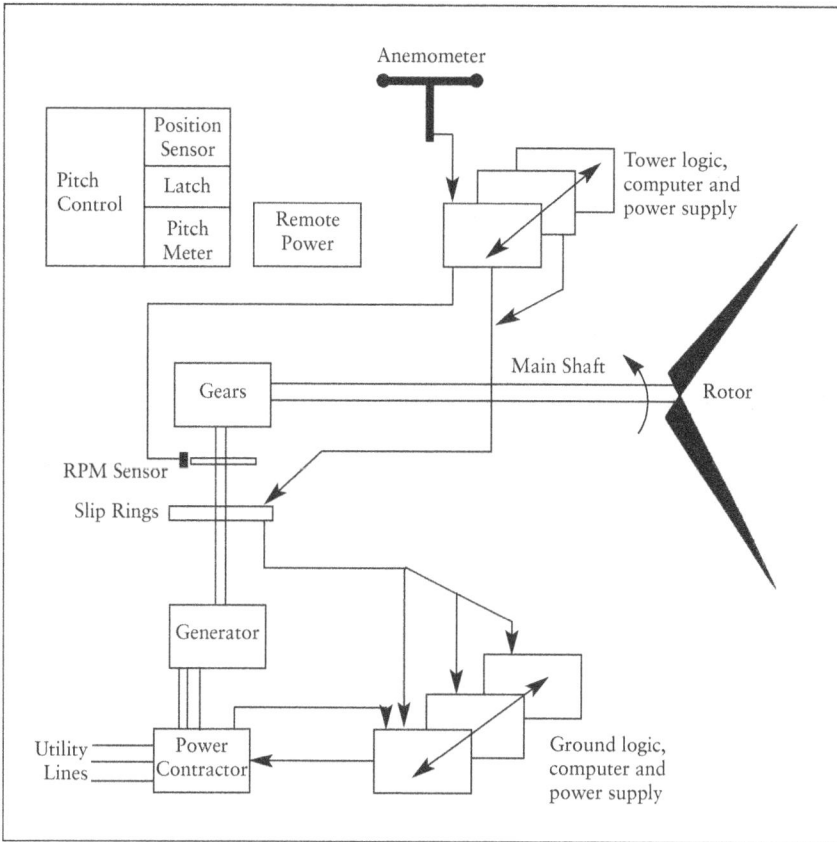

Figure 8.3   Diagram of a Typical Microprocessor Control System

- changing the intercept area by changing the rotor geometry and yawing the rotor out of the wind
- applying brakes, whether mechanical, hydraulic, air or electrical
- a microcomputer is usually part of the control system that controls both the capture and dumping of power (Figure 8.3)

Annual average wind speeds greater than 18 kilometres per hour (11 mph) or 5 metres per second are required for small wind turbines. Less wind is required for mechanical applications such as water pumping. Large wind farms require wind speeds of at least 13 miles per hour (21 km/h; six metres per second). Wind turbines range in size from 50 kW to 4 MW but the wind energy industry is now concentrating on the use of turbines larger than 200 kW. The 500–600 kW turbine is the common size in many new wind farms where the trend is to use larger turbines than before.

In order to ensure continuous power availability in independent applications, hybrid systems are recommended. These combine wind energy with diesel generators. In wind-diesel configurations, wind turbines are coupled to the diesel generators on a common electric grid, therefore displacing part of the fuel requirements. The benefits of such a combination of power sources outweigh the disadvantages characteristic of the individual power source. A wind power-diesel hybrid typically will have significantly lower operating costs than a primary diesel electric generator. State-of-the-art intelligent microprocessor-based controllers are now making such systems easy to arrange. Stand-alone wind power systems provide power only when the wind is blowing; therefore the power output is variable, unless a storage system is available. A hybrid system with a diesel generator makes storage unnecessary.

Turbine Technology

Technological developments continue to evolve rapidly over time. Compared with the traditional windmills of the nineteenth and early twentieth century, a modern power generating wind turbine is designed to generate high quality, network frequency electricity and to operate continuously for more than 20 years. During the last 20 years turbines have increased in size by a factor of 400, the cost of energy has decreased by a factor of about 10, and the industry has moved from an idealistic fringe activity towards the leading edge of conventional power generation. Also the engineering base and computational tools have developed to match the machine size and volume.

Modern wind turbines usually have three blades controlled by stall or pitch regulation and either a gearbox or a direct drive system. Variable speed is an increasingly popular option particularly because it improves compatibility with the grid. The blades are usually made from glass polyester or glass epoxy. Support structures are generally tubular steel towers which taper from their base to the nacelle at the top. Wind turbines have gradually increased in size and energy output. From units of 20–60 kW in the 1980s, their capacity has increased to more than 40,000 kW. Rotor diameters have increased to as much as 126 metres. The unit cost of turbines has also been reduced because of technical improvements and higher production volumes.

Mechanical noises in turbines have been practically eliminated and aerodynamic noise significantly reduced. Today, wind turbines are highly efficient, with less than 10 per cent thermal losses in system transmission. (Zervos, 2008) The potential offshore market for wind energy is now an important catalyst for the development of larger turbines. Although there are still many challenges, including increased costs for grid connection, maintenance, and foundation engineering, there are major advantages in the higher mean wind speeds and reduced turbulence to be found off the coastline.

The significant advances made in the design of blades have made them lighter and more efficient. Typically, increasing the size of a wind turbine's blades also increases the amount of energy that can be harnessed from the wind. However, as blades become bigger their weight places a strain on the gearbox. One of the largest blades on the market is almost 61.5 metres long; the three blades of the wind turbine cover an area almost the size of two football fields and can generate enough power for almost 5,000 households. Yet, at less than 18 tonnes, the turbine is relatively lightweight (Feltus, 2008).

Baram Engineering Limited (BEL) has developed a Vertical Axis Wind Turbine that operates primarily on the principle of magnetic levitation without the need for a gear box. The multiple blades are rotating on an "o type" ring rail and can be expended easily. Compact in size, it can work with a strong wind speed of up to 50 meters per second (m/s) and as low as 3 m/s. With a small footprint the landscape required for a 10 MW plant is only approximately 7 acres. The large number of blades, up to 240 in a system, improves efficiency. Because the generating mechanism is ground based, maintenance is more readily facilitated. As a result of the low revolutions per minute, little noise is generated. Market sources indicate that the cost of this technology is approximately US$3 million per MW installed. However, with significant market penetration, the cost may be reduced.

New ideas are also being developed in relation to wind turbine technology using a system derived from jet propulsion. This flow design has high turbine efficiency, low wind utility, and thus more universal applicability.

Integrating with the Electricity Grid

The existing installed capacity of a utility company should be a minimum of 50 MW for easy integration of wind power into its grid. Nonetheless, there may be challenges. Wind power is an intermittent energy source because it depends on a variable resource. There are variations in the wind regime on an hourly, daily, seasonal and annual basis. For this reason, excessive wind power in the energy system could reach a level where the impact would have a negative effect on the economics of total energy production. There are optimum levels of wind power that can be contributed to a system beyond which problems can be created. Therefore, the quantity of wind power that can optimally be put into the total electricity production system will be limited, especially if no storage capacity is available. In addition, the distance of the wind resource from the grid could be a limiting economic factor. As a general rule, the shorter the distance from a transmission line the more economic the wind power resources will be. In any event, a resource more than 50 kilometres from a transmission line is unlikely to be economic. A further consideration is that in certain areas, long overhead transmission lines are considered a visual blight.

Specific characteristics of the utility systems will allow the optimum input of wind energy to range from a low of 10 per cent to, in special circumstances, a high of 50 per cent. It is reasonable to conclude that where wind power input is less than 10 per cent of total electricity production, no significant problems should occur. This 10 per cent level of penetration should cause no practical economic disadvantage to accompany the growth of wind power in any country over the next 30 years because potential wind resources normally fall within the range of 8–10 per cent of a country's energy output.

Installed Capacity and Market

The countries with the highest total installed capacity are Germany (22,247 MW), USA (16,904 MW), Spain (15,145 MW), India (7,844 MW) and China (5,904 MW). Ten countries around the world can now be counted among those with over 2,000 MW of wind capacity, with France, UK and Portugal reaching this threshold in 2007. In terms of new installed capacity in 2007, the US continued to lead with the all-time record of 5,244 MW.

The European region continues to lead the market with 57,136 MW of installed capacity at the end of 2007, representing 61 per cent of the global total. In 2007, the European wind capacity grew by 19 per cent, producing approximately 120 TWh of electricity, equal to 3.8 per cent of total EU electricity consumption in an average wind year. Despite the

The world's first deep water wind farm development was installed off-shore north eastern Scotland in 2007 using 5 MW turbines

continuing growth in Europe, the general trend shows that the sector is gradually becoming less reliant on a few key markets, and other regions are starting to catch up with Europe. The growth in the European market in 2007 accounted for just 43 per cent of the total new capacity, down from nearly three quarters in 2004. For the first time in decades, more than 50 per cent of the annual wind market was outside Europe, and this trend is likely to continue into the future. While Germany and Spain still represent 60 per cent of the EU market, we are seeing a healthy trend towards less reliance on these two countries. In the EU, 3,365 MW were installed outside of Germany, Spain and Denmark in 2007. In 2002, this figure still stood at only 680 MW. The figures clearly confirm that a second wave of European countries is investing in wind energy.

The encouragement of renewable energy by governments and utility companies is stimulating market penetration. This will help wind energy compete against conventional electricity generation. One force in the wind energy market has been a public fascination with energy independence, that is, a desire to be less dependent on a central utility system. This fascination has been propelled by the 'energy crisis' and the 'do-it-yourself' hobbyist. The hobbyist learns that the aesthetic aspect of a wind farm is also important, and can be dealt with by careful landscaping and by siting clusters of machines in the most pleasing arrangements.

Wind for Electricity in the Caribbean

In the parish of Manchester, Jamaica, there is a relatively large wind farm with an initial installed capacity of 20.7 MW utilizing 23,900 KW turbines. This is the largest wind farm in the English-speaking Caribbean. This wind farm is slated for expansion by approximately 18 MW. A further 60 or more MW of wind energy could be economically produced onshore Jamaica. Areas in Jamaica with potential for wind energy include most coastal areas but particularly those of St Thomas, Clarendon, Manchester, St Elizabeth, St Ann, St Mary and Portland.

Curaçao has a successful, small, modern wind farm at Tera Kora, owned and operated by KODELA, the state utility company. The farm consists of twelve 250-kW turbines operating as a baseload plant and has been supplying electricity to the island grid since July 1993. In the first two years of operation, it had an availability factor of 94 per cent and the performance of the turbines indicated a 38 per cent capacity factor. A second phase of development added another 6 MW of wind capacity into the system. At the same time there has been a history of failure in wind energy demonstration projects in Antigua, Barbados, Barbuda and Montserrat. The failures were due to a number of factors, the most important being the inadequacy of reliable wind speed data prior to construction. Barbados, Guadeloupe and Antigua are presently planning to use wind as an energy source contributing to the national grid. Anemometric studies of the wind regime are required for at least twelve months before deciding to erect wind turbines.

Environmental and Safety Issues

Wind energy is a partial solution to many environmental problems. For example, in 1995 California's wind energy plants offset the emission of more than 2.9 billion pounds (1.3 billion kg) of carbon dioxide from relatively clean-burning gas-fired power plants. These same wind energy plants offset 16 million pounds (7.2 million kg) of nitrogen oxide, sulphur dioxide and particulates which would have been emitted from California's oil-fired power plants. Wind energy does not need water in its generation system, and has no effect on water supplies. Less than 10 per cent of the total area of a wind farm is used for harvesting wind energy, thus normal agricultural and herd-raising activities can continue, resulting in multiple land use. A landowner can often earn more from wind leases than from conventional land uses. In addition, once installed, wind turbines have little or no impact on ground-based flora or fauna.

Wind energy has negative impacts, such as its interference with electromagnetic signals. It is necessary to avoid microwave transmission paths and locate wind farms about 5 km clear of radio stations and airports. Aeroacoustic noise generation is another potential problem. Danish regulations recommend a 45 decibel (dB) level at a 200 metre distance from the nearest dweller. A further consideration is that the operation of a wind turbine can vary both in tone and direction, so that even a noise below the prescribed level can sometimes cause uneasiness.

Safety for people and property is also an environmental concern in wind farm installation. Blades or nacelle structures can fall, so proper distance clearance is required for safety. In Denmark, regulations forbid the siting of a wind farm within 90 metres plus 2.7 times the rotor diameter of roads and buildings. Recent turbine designs are much safer. Another cause for concern is that operation of wind turbines may hinder the movement of birds and even cause bird-kills, but new data suggest that this is not a major problem. However, the application of wind energy should be carefully sited in regions lying directly on migratory pathways for birds. A body of studies has shown that slower moving rotors and tubular towers are less deadly to birds (Asmus, 1995).

The opportunities and concerns cited above should be discussed with the public. To this end, the initial phases of a wind energy project should include communicating with the local public through meetings or seminars. It should also be noted that wind energy's most significant impact is visual. The wind industry will have to address the aesthetic impact if wind energy is to achieve its full potential.

A Fuel of Choice

Two simple lessons are to be put in focus when making energy plans. First, energy demand will grow with economic development. Second, energy systems have long lead times and life spans. Rapid changes are difficult and we need to plan carefully. Energy is no longer 'just another commodity', but a precondition for all commodities, a basic factor essentially equal with air, water, and earth.

Wind energy is becoming increasingly popular with utilities for reasons other than new cost competitiveness. The distributed nature of the wind resource means that generators can be placed closer to demand centres, cutting transmission losses. The modular nature of wind farms and their speed of construction permit flexibility in planning. Because the fuel is free and the wind resource seasonally predictable, wind energy costs can be anticipated with a high degree of confidence, since these costs are not exposed to fluctuations in conventional fuel price and availability.

Caribbean countries should carefully consider making wind energy a 'fuel' of choice as they diversify from fossil fuels. This choice will be influenced by factors such as availability, consumer prices, government policy, environmental standards and improvements in technology. A detailed sequence of planning and studies is required for wind energy development in a utility system (Figure 8.4). In order to convincingly promote the installation of wind power capacity as a sustainable part of the energy supply system, careful studies should be done on assessing the cost benefit of high wind energy penetration. The stabilization of emissions requires the development of competitive non-carbon energy sources. Wind energy is at present at the top of potential renewable energy sources in the Caribbean.

The wind industry is now at a stage where it is being regarded by many as mature technology and able to stand on its own commercially when compared with fossil fuels. There is still the potential for greater growth that can be fostered by continuing vigorous Research and Development (R&D) effort. But if wind energy is to be fully competitive with conventional power generating technologies, even without internalization of external costs to society or reform of the very large subsidies received, the wind energy sector must make further cost reductions. About 50 per cent of the cost reductions in the last two decades are estimated to be the result of economies of scale brought about by increased market volume, in turn a result of market volume in a few countries. The remaining 40 per cent of cost reductions can be directly attributed to research and development (Zervos et al., 2008). Upfront capital costs are especially high using wind energy technology. Recently wind turbine prices have been increasing due primarily to a demand and supply imbalance. Presently they can range from approximately US$1.5–2.5 million per MW for newly built projects. The operation and maintenance constitute

```
┌─────────────────────────────┐      ┌─────────────────────────────┐
│ Identify wind resources and │      │ Estimate electricity        │
│ estimate wind power capacity│      │ consumption over short and  │
│                             │      │ medium term                 │
└─────────────────────────────┘      └─────────────────────────────┘
              │                                    │
┌─────────────────────────────┐      ┌─────────────────────────────┐
│ Investigate wind resources  │      │ Investigate electricity     │
│ • analyse reliability of    │      │ production versus demand     │
│   wind data                 │      │ • need for capacity         │
│ • evaluate environmental    │      │   expansion                 │
│   constraints               │      │                             │
└─────────────────────────────┘      └─────────────────────────────┘
              │                                    │
      ┌──────────────────────────────────────────────────┐
      │ Wind power not suitable if grid less than 50 MW.   │
      │ Wind resources not viable if more than 50km from   │
      │ transmission lines                                 │
      └──────────────────────────────────────────────────┘
              │                                    │
┌─────────────────────────────┐      ┌─────────────────────────────┐
│ Estimate wind power         │      │ Estimate cost for electricity│
│ generating costs over time  │      │ production from conventional │
│ (5–20 years)                │      │ sources over time           │
│                             │      │ (5–20 years)                │
└─────────────────────────────┘      └─────────────────────────────┘
              │                                    │
      ┌──────────────────────────────────────────────────┐
      │ Compare wind energy costs versus conventional      │
      │ energy including the environmental cost of         │
      │ conventional energy                                │
      └──────────────────────────────────────────────────┘
                          │
      ┌──────────────────────────────────────────────────┐
      │ Make wind power generation studies considering,    │
      │ amongst others:                                    │
      │ • Status of wind resource studies                  │
      │ • Wind power input should be 10–25 per cent of     │
      │   total electricity generation                     │
      │ • Growth rate of total installed capacity          │
      │ • Environmental issues                             │
      │ • Government energy and industrial policy          │
      └──────────────────────────────────────────────────┘
                          │
      ┌──────────────────────────────────────────────────┐
      │ Create plans and implementation schedule for       │
      │ significant production of wind energy              │
      └──────────────────────────────────────────────────┘
```

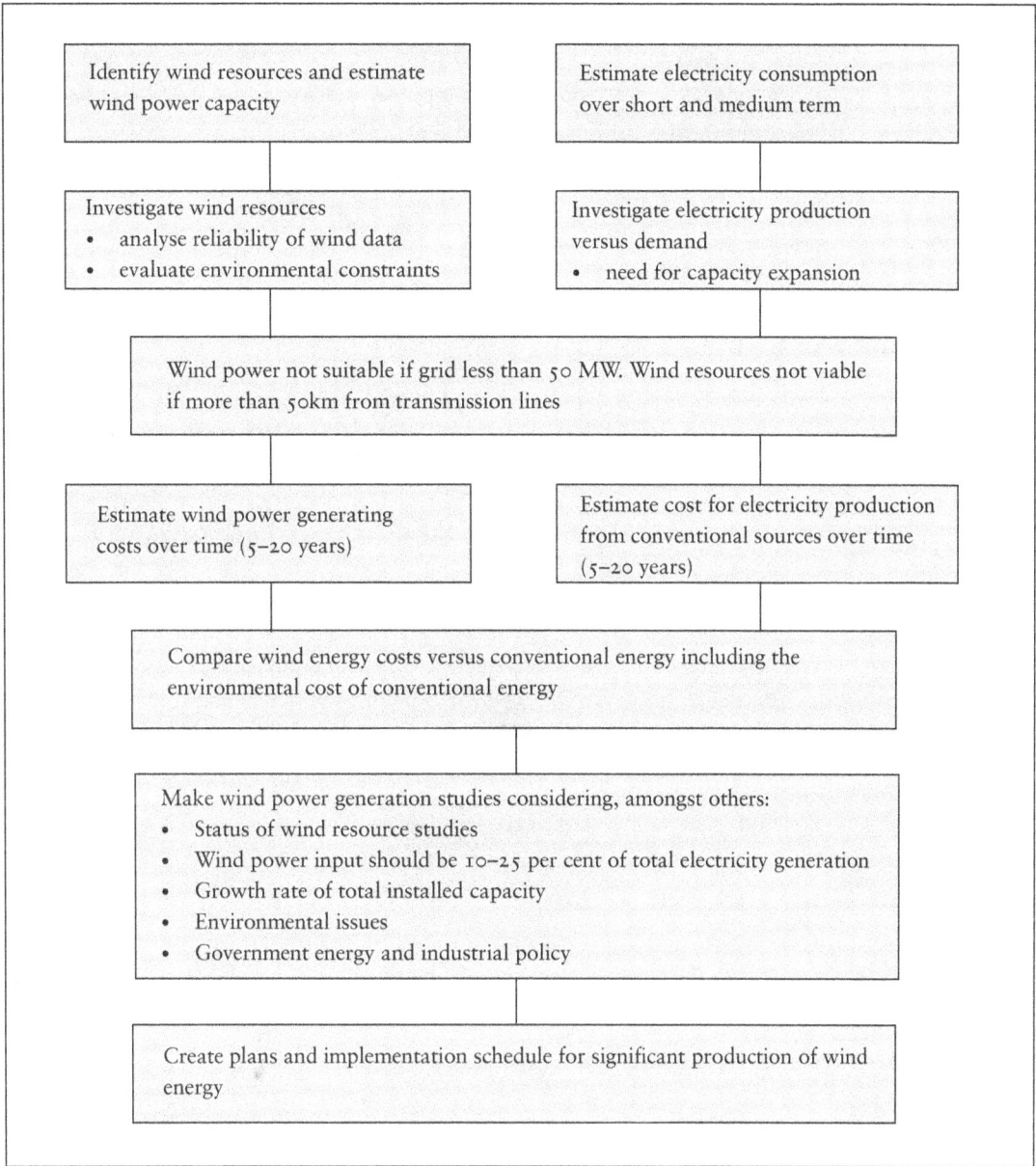

Figure 8.4 Sequence in Examining Wind Power Penetration in a Utility Company or Country

the most uncertain category because it can vary substantially from onshore to offshore models and because few operating wind turbines have achieved the end of their lifetime.

In summary, wind energy can make a major contribution towards satisfying the global need for clean renewable electricity within the next 30 years. Wind energy is one of the major options for reducing major emission reductions in the power sector, the other two being efficiency and fuel switching from coal to gas. Zervos (2008: 123) estimates that wind energy's contribution in the period up to 2020 could be reductions of more than 8 billion tonnes of CO_2. Though it may be practical to ask whether these huge numbers are achievable, it is likely that in the future governments will be asking the wind industry if more cannot be achieved.

References

Asmus, P. (1995) "Studies on Birds Offer Hope". *Windpower Monthly,* December, pp. 25–26.

Buisseret, D. (1975) "The Windmills of St Thomas". *Jamaica Journal* 9 (4) pp. 36–37.

Feltus, A. (2008) "The Blade's the Thing". *Petroleum Economist* April, p. 17.

Goodbody, I. and Thomas-Hope, E.M. (2002) *Natural Resource Management for Sustainable Development in the Caribbean.* Barbados: Canoe Press.

World Energy Council "Wind Energy: Capacity and Generation in 2007" . Retrieved from: http://www.worldenergy.org/documents/wind_11_1.pdf

Zervos A. (2008) "Status and Perspectives of Wind Energy". IPPC Scoping Meeting on Renewable Energy Sources Proceedings January 20–25, 2008.

9. | Solar Power

Because the Sun's Light is Everywhere

MANY DEVELOPING COUNTRIES HAVE FEW conventional indigenous energy resources and are vulnerable to the availability and cost of imported energy resources. Solar radiation is a primary source of renewable energy, and solar power should be considered when a country has the following characteristics:

- a high level of dependence on imported oil for energy generation
- high electricity costs due to poor economies of scale, high fuel and transportation costs
- outstanding solar resources allowing the use of photovoltaics
- a developing economy with a growing population and growing expectations
- small incremental power needs with baseload capacity additions in the range of 10–60 MW

Among the solar technologies applicable are photovoltaics, solar thermal and solar drying.

Photovoltaics

The photovoltaic (PV) process converts radiant sunlight into direct current (dc) electricity through the use of semi-conducting materials (Figure 9.1). Solar photovoltaic systems are an important part of the move towards greater use of renewable energy. It is the only technology that does not convert renewable resource energy into mechanical energy in order to generate electricity. The modern photovoltaic cell was developed in 1954 (Figure 9.2). A photovoltaic system is a complete set of inter-

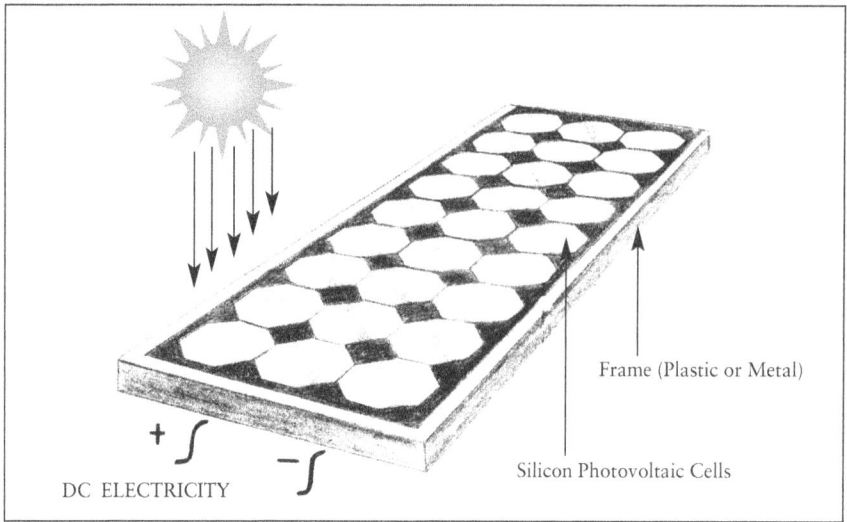

Figure 9.1 Photovoltaic Models for Generating Electricity

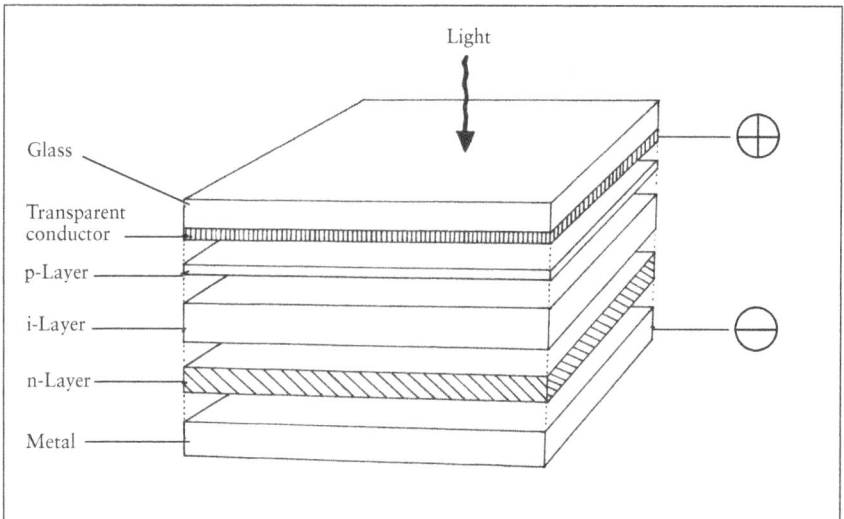

Figure 9.2 Amorphous Silicon Cell Structure

connected components that convert sunlight into electricity, including array, modules, inverters, storage batteries and the electric load. The typical photovoltaic module used for terrestrial applications contains 36 silicon solar cells, connected in series to provide enough voltage to charge a 12-volt battery. Discounted electricity production costs show that photovoltaics will continue to be more expensive than oil up to 2020 and perhaps beyond. Its application has proven technically useful in both large-scale utility networks and remote locations where other fuel sources are restricted. In addition, during the past decade, photovoltaic cell efficiencies have increased from 5 to 30 per cent.

The increasing deployment of photovoltaic technology into the user marketplace is now expected. Those working on this energy-producing technology will need to transfer it from the laboratory to the production line and build consumer confidence in PV technology. Today, the photovoltaic industry remains a dedicated yet relatively small part of the total energy production enterprise, and the goal is to develop photovoltaics for the large-scale generation of economically competitive electricity therefore making photovoltaics an important part of the world's energy mix. Notably, the annual growth rates of PV shipments worldwide have been over 20 per cent per year since 2000. This growth has been stimulated by a significant increase in venture capital for the industry.

Applications and Costs

Although the use of photovoltaics in remote applications has been increasing over time, utility-connected applications have been restricted to test facilities and prototype systems in the western USA, which are of the order of 1–6 MW. The current marketing trend is towards consumer applications such as street and walkway lights, fixed and portable lanterns, domestic lighting systems, calculators, battery charging, water pumping, refrigerators, cathodic protection, and other low power uses. The United States, Japan and Germany are meeting this market with thin-film amorphous silicon technology. When flexible PV shingles become a generally used roofing material in the next decade, houses can become self-sufficient in energy by providing their own requirements independently of the utility company. The integration of PV cells into the facade elements of buildings heralds many possibilities for the application of photovoltaics, as present projects demonstrate in Germany and in the UK (Chehab 1994; Pearsall et al., 1994).

Solar street lamps now have economic application in most low latitude countries

Photovoltaics can be used for rainscreen overcladding, roof systems, shading and curtain walling, thus enhancing its possibilities as an urban source of energy (Scott, 1994). PV installations have been used to run reverse osmosis desalination plants such as those in Frederiksted, St Croix in the US Virgin Islands. Reverse osmosis plants operate on the principle that if salt water and fresh water are separated by a semi-permeable membrane, fresh water can be squeezed out of salt or brackish water by applying pressure to the salt water side to reverse the natural tendency of the fresh water to flow by osmosis into the salt water.

Independent power producers should be able to enter into the utility market when PV systems can produce energy for about 12 US cents/kWh. Current cost estimates for on-grid applications are about 18 US cents/kWh (Ahmed, 1994). The PV module will represent about one-half of this cost. The ultimate goal is for photovoltaics to produce energy for 6–9 US cents/kWh. At these low prices, which are not expected to occur until 2020 or thereafter, photovoltaics could provide up to 8 per cent of the electricity needs of tropical countries such as Jamaica by the year 2025.

The high initial cost is a hurdle. To address this, financing is required to spread the high capital cost of PV systems over the life of the system and to make them more accessible to cash-poor potential users. Adding to costs is the fact that batteries need regular maintenance and eventual replacement. Storage batteries have a short lifetime of about 3–5 years. This cost mainly affects remote systems. Utility grid-attached systems for peak power, or PV systems used with existing hydropower schemes in a hybrid arrangement have little need for storage. In remote areas, hybrid PV-diesel generator systems typically will cost less than a stand-alone PV system for loads ranging from 6 to 250 kWh/day. Also, a PV-diesel hybrid will have a lower operating cost than a primary diesel electricity generator.

PV modules produce dc electricity only; an inverter must be added to the system to run ac devices. Although this adds to the cost of the system, it enables the use of more readily available, and cheaper, ac appliances. In addition, there are the costs associated with security: the modular characteristics that allow easy expansion of photovoltaic systems leave them vulnerable to theft and vandalism. Training is also required for the users to ensure the safe and secure operation of a complete PV system, adding to the overall cost.

Technology

Figure 9.3 shows the technology elements in a photovoltaic system. The solar cells are solid-state devices. Photons striking a solar cell are reflected or else absorbed by the cell and pass through it. The absorbed photons transfer energy to the cell. Several PV cells are interconnected in series, encapsulated and sealed, most with a tempered glass cover and a plastic support backing sheet, to form a photovoltaic module. The laminated module protects the electrical circuits from the environment and gives the long life for which photovoltaic modules are noted. Modules may be connected in series to obtain the required system voltages or in parallel to obtain larger currents. The module arrangement makes it possible to design systems to meet a variety of electric loads. All systems include wire, connectors, switches, and electrical protective components. If the load requires alternating current (ac), an inverter is used to convert the dc power to ac power.

Trackers are mechanical devices that follow the sun on its daily movement across the sky; mounting photovoltaic modules on a tracker keeps them directly facing the sun (Figure 9.4) so that they will collect more sunlight and produce more electricity than if

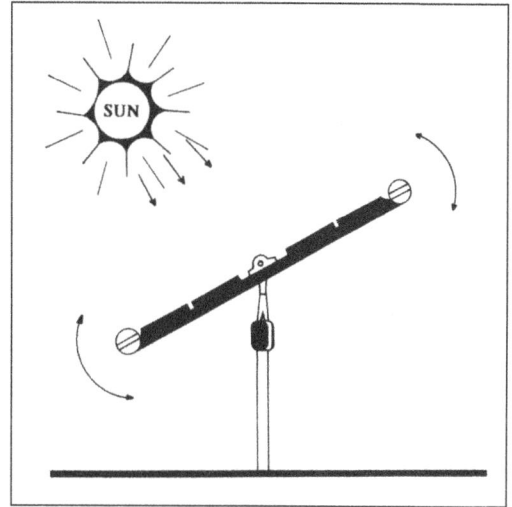

Figure 9.4 Diagram of a Mechanical Tracker that Follows the Sun in its Daily Movement Across the Sky
Source: Shepperd and Richards, 1993

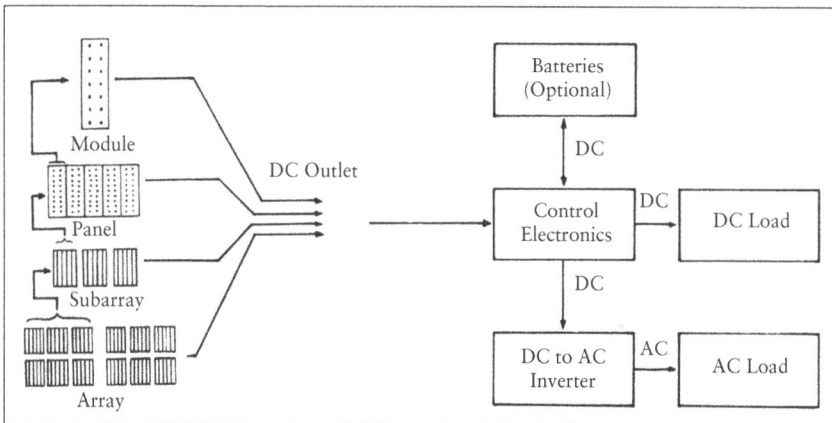

Figure 9.3 Schematics of the Production of Electricity from a Photovoltaic System

they were mounted in a fixed position. The decision to use a tracker is based on the trade-off between the cost of the tracker and the savings realized by using fewer modules to obtain a given amount of power.

PV modules are constructed from several solar cells; the modules vary in size but are generally about one square metre and deliver no more than 60–150 watts at that size. The modules are arranged in arrays which may track the sun's movement and maximize performance. The semi-conductor materials used include silicon (Si), copper indium diselenide (CIS), cadmium telluride (CdTe) and gallium arsenide (GaAs). In broad terms, there are three types of PV cells, 'thick film', 'thin film' and concentrator designs.

'Thick film' cells are made primarily from single-crystal silicon, amorphous silicon, semicrystalline and polycrystalline silicon. Thick films using amorphous silicon and polycrystalline are the dominant technologies. Silicon is abundant, forming approximately 20 per cent of the Earth's crust but the material used in thick film photovoltaic cells is usually waste material from the semi-conductor industry. Production efficiencies are approximately 12–16 per cent whilst experimental efficiencies are of 20–25 per cent. Theoretical efficiencies for single-crystal silicon cells are just over 30 per cent. Polycrystalline silicon cells are less efficient than single-silicon cells, with efficiencies of 8 per cent in production modules. They have applications in less energy-intensive processes. Recent improvements have increased the performance of bulk polycrystalline Si cells to the 17 per cent level, according to Kazmerski and Emery (1994) who give a thorough review of these technologies.

'Thin film' cells require much less material than single-crystal silicon. The films are typically between 0.001 and 0.002 mm thick, compared with 0.3 mm for a typical thick film silicon cell. Because the manufacturing techniques are less costly than making ('growing') single cells, they have much potential for low-cost, high-volume production. These lower efficiencies, which are not always stable, are countered by lower cost per unit area because smaller quantities of active material and lower temperatures are needed in production. Significant improvements have been made in the polycrystalline technologies using two thin film options, CIS and CdTe. Where CdTe is used, production efficiencies are of the order of 10 per cent and can now be used on a commercial scale. CIS cells have similar production efficiencies of about 11 per cent. When efficiencies reach the goal of 15 per cent, these polycrystalline technologies could become the focus of PV use.

A further development is the concentrator cell which is an attempt to reduce costs. Mirrors or lenses are used to concentrate the sunlight into a photovoltaic cell with a smaller area. At present, single-crystal gallium arsenide and single-crystal silicon are used in concentrator cells which require direct sunlight and competent tracking mechanisms. Production efficiencies range up to 25 per cent, with theoretical efficiencies reaching as high as 33 per cent. Other advanced approaches include an electrolytic cell, the gratzel cell, which uses a spectrally sensitized, thin ceramic membrane. However, performance is not stable and much more work has to be done to establish a path to its commercialization.

Limitations

There are several limitations to photovoltaic energy. Firstly, an engine generator will not run without fuel, and without sunlight photovoltaic systems cannot run. Sunlight is a diffuse fuel source, hence PV systems are energy limited. With the present state of technology, photovoltaic systems are probably not the optimum choice for applications with high-power requirements, such as air conditioning, especially if these applications are needed at night. Battery storage is required for night use, increasing the complexity of the system.

Impact on the Environment

In the manufacturing process, only negligible quantities of carbon dioxide are released into the atmosphere during the reaction of silica with hydrogen to form silicon and carbon dioxide. Silicon does not pose an environmental health hazard on disposal. This is not necessarily the case with other heavy metals such as cadmium telluride, cadmium sulphide, gallium arsenide and copper indium diselenide, all of which can be hazardous if the solar cell arrays are burnt since they will produce poisons (arsenic) or toxic gases. Cadmium, for example, is poisonous and possibly carcinogenic, making it a health risk both at the production and disposal phase. Good factory management is essential. Table 9.1 shows the potential environmental hazards of semi-conductor materials and of cadmium telluride in particular. The data suggests that cadmium (Cd) and its compounds in the PV industry should be treated with care in order to minimize detrimental effects on health and the environment.

Table 9.1   Potential Environmental and Health Hazards of Semiconductor Materials

Material	Potential Environmental and Health Hazards
Cd	• Spontaneous flammable in air • Can react with some acids and oxidizing agents • Reacts with sulphur, selenium, tellurium and zinc • Violent reaction with potassium. Insoluble in water • Prolonged inhalation of fumes or dust can produce systemic effects • (TCLo 88 µg/m3/8.6 years-inh.man.) • May be harmful if ingested
CdS	• May evolve toxic fumes in fire • In reaction with water, hydrogen sulphide may be evolved • Reacts with acid to produce hydrogen sulphide • Harmful by ingestion, inhalation and skin contact • Practically insoluble in water. Irritating to eyes • Can cause increased salivation, choking, vomiting, stomach pains and diarrhea • Prolonged exposure to dust can cause lung and kidney damage • Contact with gastric juices releases toxic hydrogen sulphide.
$CdSO_4$	• May evolve toxic fumes in fire • No reaction with other materials. Very soluble in water • Harmful by ingestion, inhalation and skin contract • Irritating to eyes • Can cause increased salivation, choking, vomiting, stomach pains and diarrhea • Prolonged exposure to dust can cause lung and kidney damage • Danger of cumulative and irreversible effects
CdTe	• May evolve toxic fumes in fire • In reaction with water, hydrogen telluride may be evolved • Reacts with acids producing hydrogen telluride • Practically insoluble in water • Harmful by ingestion, inhalation and skin contact • Irritating to eyes • Can cause increased salivation, choking, vomiting, stomach pains and diarrhea • Prolonged exposure to dust can cause lung and kidney damage • Brief exposure to high concentration may result in pulmonary oedema and death

Table 9.1 continues

Material	Potential Environmental and Health Hazards
Te powder	Combustible • May evolve toxic fumes in fire • Can react violently with metals, halogens and inter-halogen compounds • Immiscible or insoluble in water • Harmful by ingestion and inhalation causing nausea, vomiting, central nervous system depression and garlic odour on breath. • Irritant to skin and eyes
TeO_2	• May evolve toxic fumes in fire • Practically insoluble in water • Harmful by ingestion and inhalation causing nausea, vomiting, central nervous system depression and garlic odour on breath • Irritant to skin and eyes

Source: Baumann et al., 1995

Photovoltaic field applications require a large land area (0.8 km²/per 100 MW) but grass and other low-lying growth remain largely undisturbed. Small livestock can therefore graze in these dual-purpose areas.

Solar Thermal

Solar thermal systems collect the thermal energy in solar radiation for use in low to high temperature thermal applications. Several types of collection systems (parabolic trough, central receiver and parabolic dish) may be used to concentrate and convert the solar resource.

The leading solar thermal technology is the parabolic trough, which focuses the sunlight on a tube which carries a heat-absorbing fluid, usually oil. The fluid is circulated through the line focus tubes of the parabolic trough collectors, reaching temperatures up to 391°C, and is used to boil water to steam which is then routed to a high-pressure turbine to generate electricity. Turbine efficiency is almost 38 per cent. Approximately 360 MW of parabolic trough electricity generating capacity is now operating in California's Mojave Desert, mostly to provide peak power during the high-demand periods which occur during the summer afternoons. The overall efficiency of these systems is almost 23 per cent, higher than the efficiencies that obtain for PV cells. This single solar thermal installation is nearly as large as the combined PV installations worldwide. At 8 US cents per kWh, the electricity cost is much lower than from PV. The major drawback is the large amount of land required for a solar thermal power plant. One hectare of land is used to produce approximately 1.6 MW of power. The technology is not yet suitable for small island states.[1]

Solar towers are a type of solar furnace using a tower to receive the focused sunlight. It uses an array of flat, movable mirrors as heliostats to focus the sun's rays upon a collector tower. The high energy at this point of concentrated sunlight is transferred to a substance that can store the heat for later use. The most recent heat transfer agent that has been successfully demonstrated is liquid sodium. Sodium has a high heat capacity, allowing energy to be stored and drawn off as required at nighttime. Further, that energy can be used to boil water for use in steam turbines. In an initial power tower version located at Solar One in Barstow California, water was used as a heat transfer. This system did not allow for power generation when the sun was not shining.

Power towers enjoy the benefits of two successful, large-scale demonstration plants. The 10-MW Solar One plant near Barstow, California, demonstrated the viability of power towers, producing over 38 million KWh of electricity during its operation from 1982 to 1988. The Solar Two plant was a retrofit of Solar One to demonstrate the advantages of molten salt for heat transfer and thermal storage. Utilizing its highly efficient molten-salt energy storage system, Solar Two successfully demonstrated efficient collection of solar energy and dispatch of electricity, including the ability to routinely produce electricity during cloudy weather and at night. In one demonstration, it delivered power to the grid 24 hours per day for nearly 7 straight days before cloudy weather interrupted operation.

The successful conclusion of Solar Two sparked worldwide interest in power towers. As Solar Two completed operations, an international consortium, led by US industry including Bechtel and Boeing (with technical support from Sandia National Laboratories), was formed to pursue power tower plants worldwide, the largest plant in Spain (where special solar premiums make the technology cost-effective), and other plants located in Egypt, Italy, Morocco and other countries. The first commercial power tower plant of the consortium is planned to be four times the size of Solar Two (about 40 MW equivalent, utilizing storage to power a 15 MW turbine up to 24 hours per day).

Parabolic dishes are useful for small-scale applications in remote locations. A parabolic dish tracks the sun and focuses its heat onto a Stirling engine, which converts the heat energy· to mechanical energy. The mechanical energy drives a turbine to generate power. Such systems can generate 5–25 kW of power.

The cost of energy has decreased for solar thermal, in part not only because of economies of scale of manufacturing and deployment of more systems, but also because of improved systems as a result of successes in research and development.

Solar Water Heating

Solar water heating systems are the primary solar application in the Caribbean. In 1995 there were just under 2,000 solar water heaters installed in Jamaica, in stark contrast to Barbados which has a population ten times smaller than Jamaica. Triggered by economic incentives in the 1980s, there are approximately 70,000 solar water heaters installed in Barbados. The Jamaican government has taken steps to promote solar water heaters by having the solar panels zero rated under General Consumption Tax. Although the success story of Barbados may not be replicable in all the CARICOM countries, the solar water-heating industry should experience a significant expansion in Jamaica over the next five years, resulting in savings on imported oil. Husbands (1995) indicates that in one year, 1992, Barbados effected savings of US$9.75 million from the 23,388 units installed with a total capacity of 1,403,725 gallons. In the past decade, an important part of the electricity supply has been used for purposes other than the heating of water.

Solar water-heating technology is reliable and the equipment is relatively simple to manufacture and install. Most installations have a payback period of four to six years. Although hot water may be considered a luxury in some Caribbean countries, there is still a large demand for hot water in households and in the tourist industry. Solar water heaters should be used for preheating water in all hotels before raising the temperature by electrical methods. Also, heat exchangers could be connected to the central air-conditioners so that the heat removed from guest rooms could be used to assist the heating of water (Headley, 1994). Eventually, the use of solar water heaters may be made mandatory for certain uses in the Caribbean. Egypt and Western Australia require all government buildings to have solar water heaters while new houses in Israel must have solar water heaters.

Solar Drying and Distillation

Applications for solar drying and distillation include but are not limited to crops, timber and the desalination of water. Even though the potential for solar crop drying has not been fully realized, it is a means of preventing spoilage which affects as much as 30 per cent of crop production in tropical countries. Crops such as bananas, papaya, sorrel, sweet potato, yam, ginger, nutmeg, pimento, grasses and leaves can be dried by solar dryers that range from the simple wire basket dryer of approximately two square metres to roof solar collectors (Headley, op. cit). Table 9.2 shows the general characteristics and performance of some simple solar dryers.[2]

When timber is sun-dried, few devices are used to deliberately speed up the process. Purpose-built devices are approximately five times faster

Table 9.2   General Characteristics of Selected Simple Solar Crop Dryers

Dryer Characteristic	Type A	Type B	Type C	Type D
Collector Area m^2	3.3	2.8	7.8	2.2
Collector Slope	20°	15°	15°	10°
Collector Plate material	Finned sheet metal	Sheet metal with vanes	Sheet metal	Wire mesh
Transparent cover	Glass, double glazed	Glass, single glazed	Glass, double glazed	Plastic sheet
Loading density, kg/m^2	20	10	20	10
Normal drying Temperature (^{o}C)	65	40	50	40
Typical drying time, hours	15	20	20	20
Materials cost, US$/m^2	106	95	180	27
Labour cost, US$/m^2	62	62	125	10

Source: Headley, 2005

than sun-drying or air-drying and can bring products down to 10–16 per cent moisture content within three days (crops) or twenty-one days (timber). Timbers such as mahogany, oak and teak which are used for high-quality furniture, need to be dried to a moisture content of 10 per cent, consistent with the stability of the finished furniture.

The traditional method of drying in ambient air takes three or more months, involving the payment of interest on capital, and the costs of the conventional hot air kilns which, when used, are expensive to purchase and operate (Headley, op. cit.). In a large timber-producing country, such as Guyana, drying could be carried out by means of solar energy or by using timber residues to fuel a hot air kiln. Solar distillation units produce relatively small volumes at a rate of about 3–6 litres per square metre per day. Approximately fifty solar stills have been installed in Trinidad and Tobago as well as Jamaica but not all are currently in operation. They are chiefly 2.2 square metre units which produce 6 to 13 litres of distilled water per day. Solar stills can be used for the desalination of sea water but are not economic when compared with other desalination techniques such as reverse osmosis. Thus, solar distillation is used primarily for producing distilled water for science laboratories and for use in automobile batteries and appliances which require dematerialized water.

Future Prospects

The solar photovoltaic market has three major segments: consumer products, remote power and utility generation. Although the consumer market was once dominated by calculators and watches, battery charging and security lighting are growing rapidly and have become the largest part of this market. Utility companies are investigating the technology both for their own use and to learn how it interacts with their systems. One approach is for utilities to sell or lease small systems for remote cabins and homes. The remote power sector is by far the largest market today, providing vaccine refrigerators in health clinics, pumping and disinfecting water, providing lighting and power communication systems. French Polynesia alone has over 2,500 individual photovoltaic power systems. An important segment of the market in Jamaica provides a backup for computer systems and gives protection from the vagaries of blackouts on the grid-line electricity provided by the utility. A number of photovoltaic lighting systems have been installed for practical and demonstration purposes (Figure 9.5), and this number is set to increase during the next

Photovoltaic modules mounted on trackers which light the grounds of Jamaica House, Office of the Prime Minister, Kingston, Jamaica

decade as the utility power market comes within economic reach. The future successful use of photovoltaics in electricity grid operations awaits only the required cost reduction in solar collectors to be economical.

In order to accelerate the transition towards the use of solar energy and make important progress towards a sustainable economy, greater emphasis will be placed on commercialization and marketing so that higher targets can be achieved as the technology matures. As part of a diversified energy system, the use of photovoltaics will give lower supply risks, become more economical in the long term and help to foster new employment. Other interesting advancements that suggest a positive future for solar energy are:

- Innovative financial systems have become better dispersed. Companies that sell electricity rather than physical solar systems have emerged. The companies own the system, are responsible for maintenance whilst customers benefit from a known and predictable price for electricity from the renewable energy source. This business model facilitates access to solar energy by consumers without the outlay of the high capital cost.

- Thin film advances have created potential new applications for solar technology. For instance, the use of pigmented solar components may expand and enhance the use of solar products for commercial building applications.

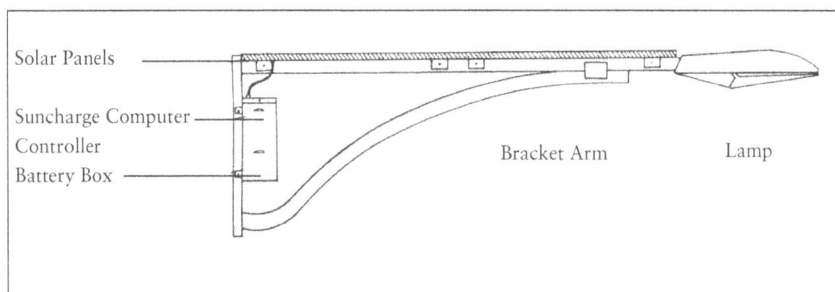

Solar Panels

Suncharge Computer
Controller
Battery Box

Bracket Arm   Lamp

Figure 9.5   Diagram of a Solar Street Lamp as Mounted at Many Locations

- Increasingly, solar technology is being used for small power requirements. Companies are expanding the market for applications such as powering electronics. This is a particularly significant development for countries where technology is widely distributed and access to an electrical power infrastructure is lacking (Arvizu, 2008).

- Passive solar concepts in building design use natural heating, cooling and lighting to reduce energy use. This low-energy architecture will be the focus of building approaches over the next two decades.

Solar energy has great potential, and developing countries of the tropical belt share one adequate resource, abundant sunshine. The use of solar energy should be incorporated in the science curriculum in schools all over the world, so that young persons gain an early appreciation of this enormous resource. Undoubtedly, solar energy will be a subject of much cooperative development in tomorrow's world.

References

Ahmed, K. (1994) "Renewable Energy Technologies". World Bank Technical Paper 240. American Solar Energy Society, 1989, *Assessment of Solar Energy Technologies*. Boulder, Colorado.

Arvizu, D. (2008) "Potential Role and Contribution of Direct Solar Energy to the Mitigation of Climate Change". National Renewable Energy Laboratory, Colorado USA, IPCC Scoping Meeting on Renewable Energy Sources Proceedings, Lubeck, Germany.

Baumann, A.E., Hynes, K.M., Hill R. (1995) "An Investigation of Cadmium Telluride Thin Film PV Modules by Impact Pathway Analysis". *Renewable Energy 6* (5–6) pp. 593–599.

Chehab, D. (1994) "The Intelligent Facade: Photovoltaic and Architecture" *Renewable Energy 5* (pt. I) pp. 188–204.

Headley, O. (1994) "Solar and Alternative Energy in the Caribbean – Prospect and Retrospective". Proceedings of meeting on Sustainable Alternatives for Tropical Island States, March 14–16, 1994, Barbados. UWI Centre for Environment and Development, pp. 45–64.

Headley, O. (1995) "Solar Thermal Systems for Use in the Caribbean". Proceedings of the Caribbean High Level Workshop on Renewable Energy Technologies December 5–9, 1994, Saint Lucia. UNESCO.

Husbands, J. (1995) "A Review of the Costs of the Tax Incentives to the Solar Industry in Barbados". Proceedings of the Caribbean High Level Workshop on Renewable Energy Technologies, December 5–9, 1994, Saint Lucia. UNESCO.

Kazmerski, L.L., and Emery, K. A. (1994) "Photovoltaic Technologies: Evaluation and prospects". Proceedings of meeting on Sustainable Alternatives for Tropical Island States, March 14–16, 1994, Barbados. UWI Centre for Environment and Development, pp. 45–64.

Pearsall, N.M., Hill. R., Claiden, P. (1994) "PV Cladding as an Energy Resource for the UK". *Renewable Energy* 5 (pt. I) pp. 348–355.

Scott, R.D.W. (1994) "The Future of Solar Power in the Urban Environment". *Renewable Energy* 5 (pt. L) pp. 225–228.

Shepperd, L.W., and Richards, E. (1993) "Solar Photovoltaics for Development Applications". Scandia National Laboratories SAND 93–1642.

10. | Bioenergy

An Infinite Fuel Source

RECENTLY THERE HAS BEEN A movement towards the greater use of bioenergy in line with the concept that countries should grow as much fuel as possible. Bioenergy is a form of energy derived from living or recently living organisms or their metabolic byproducts. Biomass is the major source of bioenergy. Biomass may be defined as any organic substance other than oil, natural gas and coal. It includes trees, agricultural crops, organic waste, residue and effluent of agricultural, agro-industrial and domestic origin. Biomass has been in use for many years, mainly in the form of agricultural and industrial waste which is burned to fuel conventional steam turbines to produce electricity. Biomass accounts for about 15 per cent of world energy use and 36 per cent of energy use in developing countries. In Jamaica, biomass is the source of approximately 11.5 per cent of energy use.

The Formation of Bioenergy

Plant materials use the sun's energy to convert atmospheric carbon dioxide to sugars during photosynthesis. Energy is released on combustion of the biomass as the sugars are converted back to carbon dioxide. In this manner, biomass serves as a form of natural battery for storing solar energy. The fact that the energy is marshalled and released in a short time span makes biomass a renewable and sustainable energy source. The maximum theoretical value of biomass is 6.7 per cent for C4 plants (the first product of photosynthesis is a 4-carbon sugar) such as corn, sorghum and sugar cane which thrive in the tropics. In other words, these crops have a very large potential for producing biomass.[1] Plants such as

Trees such as these in the Caledonian Forest of Scotland are useful for removing carbon dioxide from the atmosphere

wheat, rice and most trees which are identified as C3 have an efficiency value of 3.3 and account for 95 per cent of plant biomass globally (Ahmed, 1994). Factors such as disease, pests, leaf cover, temperature, water and nutrients make these values lower in the range of 2 to 3 per cent for C4 and about 1 per cent for C3 (Hall et al., 1993). These photosynthetic efficiencies point to a high land intensity for biomass energy when compared with other sources of energy such as photovoltaics, which have a solar energy to electricity conversion percentage of 3 to 17 per cent in the field, experimental efficiencies of 6 to 34 per cent and theoretical efficiencies of 47 per cent (Ahmed, 1994). These percentages indicate that arable land might be better used for crop production.

Factors Related to Production

As illustrated by Figure 10.1, biomass can be converted to energy by:

- direct combustion with the heat used for cooking and space heating
- combustion to generate steam which is used to drive a turbine
- biochemical or thermochemical degradation of biomass to form biogas and liquid fuels

Biogas production is obtained through anaerobic digestion, which is

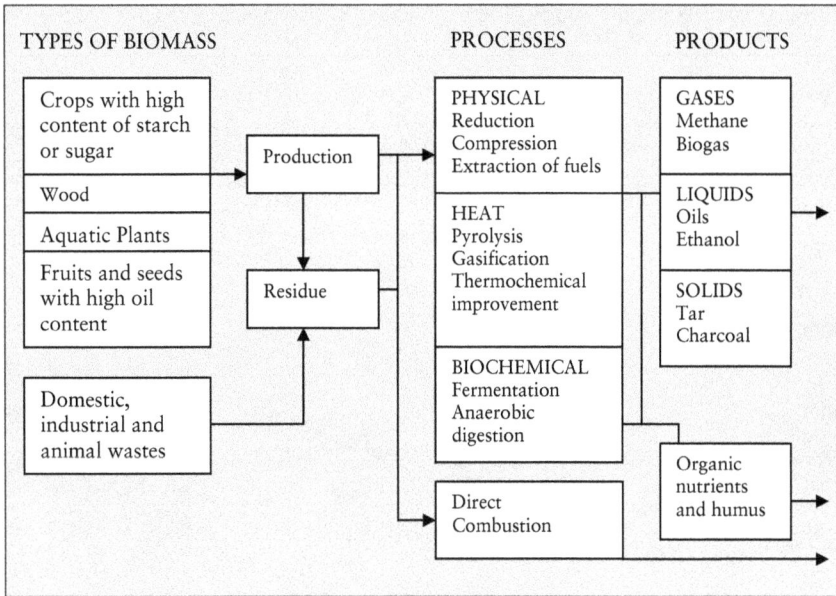

Figure 10.1 General Model of Biomass Production

a biological process that converts solid or liquid biomass into gas in the absence of oxygen. The gas consists mainly of methane and carbon dioxide and contains various trace elements.

Liquid Fuels

The fermentation of sugars to produce ethanol is an old process which produces alcohol, ethanol, and methanol.

Ethanol may be formed from sugars (sugar cane), starches (corn and cassava) or cellulosic material. Where sugar cane is concerned, the cane is directly fermented to form ethanol with the waste bagasse being burned for cogeneration. In the case of starches and cellulosic material, they must first be broken down into sugars before fermentation, using acids or hydrolytic enzymes.

The major source of ethanol is molasses, a by-product of sugar production. The annual world production of molasses is about 33.5 x 106 tonnes sold and 5 x 106 tonnes unused. Molasses contains between 46 and 50 per cent fermentable sugar which can be converted to ethanol according to the following reaction:

$$C_6H_{12}O_6 \longrightarrow 2C_2H_5COOH + CO_2$$

Table 10.1 Average Ethanol Yields of Various Biomass Sources

Source	Yields (Litres ethanol/tonne biomass)
Molasses	270
Sugar Cane	70
Sweet Sorghum	80
Corn	370
Sweet Potato	125
Cassava	180
Babassu	80
Wood	160

Source: Luo, 1994

Table 10.1 shows the average ethanol yields per ton from different biomass sources using existing technologies. Corn and molasses are the most efficient sources.

Blending methanol or ethanol with conventional fuels is a low-cost opportunity of introducing renewable resources such as biomass (and also wastes) to the fuel market. In Germany today up to 3 per cent methanol is permitted in conventional fuels by legislation. Providing this 3 per cent through renewable resources would be sufficient to initiate a new industry, setting up production capacities of converting waste wood, biomass and residues to methanol or synthetic diesel.

Methanol synthesis is a two-stage process and it is cheaper to produce methanol from natural gas or coal than from biomass. The gas can now be made by a thermochemical degradation reaction in the presence of oxygen followed by a shift-gas reaction to obtain a precise mixture of hydrogen and carbon monoxide, which, after passing through a pressurized catalytic reactor, liquid methanol is formed.

Both ethanol and methanol are mainly used as an additive to gasoline. Pure ethanol can be used in anhydrous form as a blend with gasoline or, less commonly, in its hydrated form (95:5 ratio of ethanol to water) as a transport fuel. The latter, though cheaper because it does not require the extra step of distilling the hydrated ethanol, requires special adaptation of the engines used.

Gaseous fuels are also important: for example, manure can produce biogas in digesters. An operating temperature of close to 35°C is best for digesters and a higher or lower temperature will result in less gas production, among other problems. A number of European countries are involved in a long project entitled the LEBEN Project which supports research on new technologies that upgrade biomass to solid, liquid and gaseous fuels. A central system developed in this project has been very effective (Sayigh, 1995).

The production of ethanol from starch or sugar-based feedstocks is among man's earliest ventures into value-added processing. While the basic steps remain the same, the process has been considerably refined in recent years and has become an efficient process. There are two production processes, namely wet milling and dry milling. The main difference between the two processes is in the initial treatment of the grain.

Figure 10.2 illustrates the dry milling process. In dry milling, the entire corn kernel or other starchy grain is first ground into flour, which is referred to in the industry as "meal" and processed without separating out the various component parts of the grain. The meal is slurried with water to form a "mash." Enzymes are added to the mash to convert the starch to dextrose, a simple sugar. Ammonia is added for pH control and as a nutrient to the yeast. The mash is processed in a high-temperature cooker to reduce bacteria levels ahead of fermentation. The mash is cooled and transferred to fermenters where yeast is added and the conversion of sugar to ethanol and carbon dioxide (CO_2) begins.

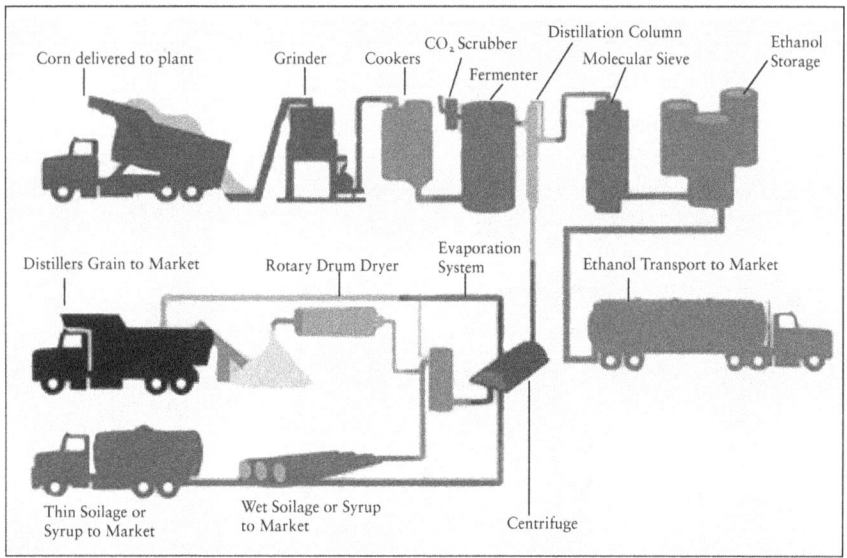

Figure 10.2 The Ethanol Production Process: Dry Milling
Source: Renewable Fuels Association

The fermentation process generally takes about 40 to 50 hours. During this part of the process, the mash is agitated and kept cool to facilitate the activity of the yeast. After fermentation, the resulting "beer" is transferred to distillation columns where the ethanol is separated from the remaining "stillage." The ethanol is concentrated to 190 proof using conventional distillation and then is dehydrated to approximately 200 proof in a molecular sieve system. The anhydrous ethanol is then blended with about 5 per cent denaturant (such as natural gasoline) to render it undrinkable and thus not subject to beverage alcohol tax. It is then ready for shipment to gasoline terminals or retailers.

The stillage is sent through a centrifuge that separates the coarse grain from the soluble. The soluble is subsequently concentrated to about 30 per cent solids by evaporation, resulting in Condensed Distillers Soluble (CDS) or "syrup." The coarse grain and the syrup are then dried together to produce dried distillers grains with soluble (DDGS), a high quality, nutritious livestock feed. The CO_2 released during fermentation is captured and sold for use in carbonating soft drinks and beverages and the manufacture of dry ice.

Figure 10.3 illustrates the wet milling process. In wet milling, the grain is soaked or "steeped" in water and dilute sulfurous acid for 24 to 48 hours. This steeping facilitates the separation of the grain into its many component parts. After steeping, the corn slurry is processed through a series of grinders to separate the corn germ. The corn oil from the germ

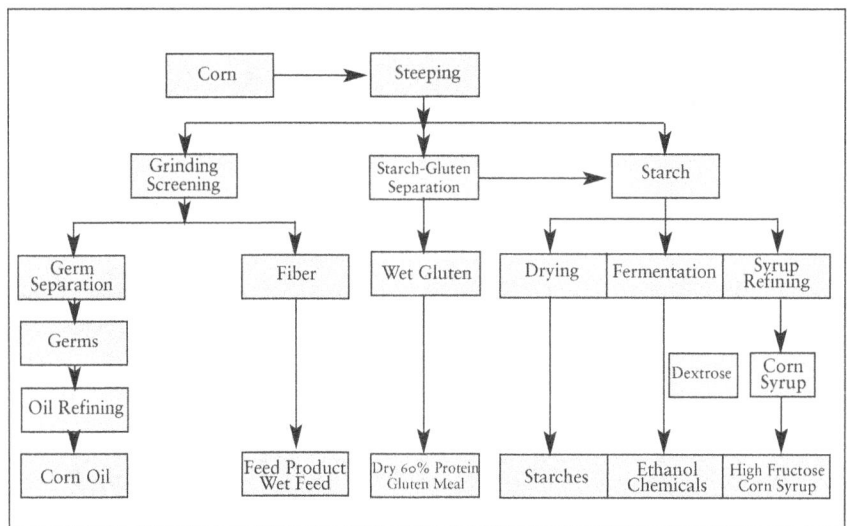

Figure 10.3 The Ethanol Production Process: Wet Milling
Source: Renewable Fuels Association

is either extracted on-site or sold to crushers who extract the corn oil. The remaining fiber, gluten and starch components are further segregated using centrifugal, screen and hydrochloric separators.

The steeping liquor is concentrated in an evaporator. This concentrated product, heavy steep water, is co-dried with the fiber component and is then sold as corn gluten feed to the livestock industry. Heavy steep water is also sold by itself as a feed ingredient and is used as a component in Ice Ban, an environmentally friendly alternative to salt for removing ice from roads.

The gluten component (protein) is filtered and dried to produce the corn gluten meal co-product. This product is highly sought after as a feed ingredient in poultry broiler operations. The starch and any remaining water from the mash can then be processed in one of three ways: fermented into ethanol, dried and sold as dried or modified corn starch, or processed into corn syrup. The fermentation process for ethanol is very similar to the dry mill process described above.

Figure 10.4 is a process flow diagram which shows the basic steps in the production of ethanol from cellulosic biomass. There are a variety of options for pretreatment and other steps in the process and that several technologies combine two or all three of the hydrolysis and fermentation steps within the shaded box.

Cellulose, hemicellulose and lignin are the initial components that provide a source of cellulosic ethanol. Cellulosic ethanol is chemically

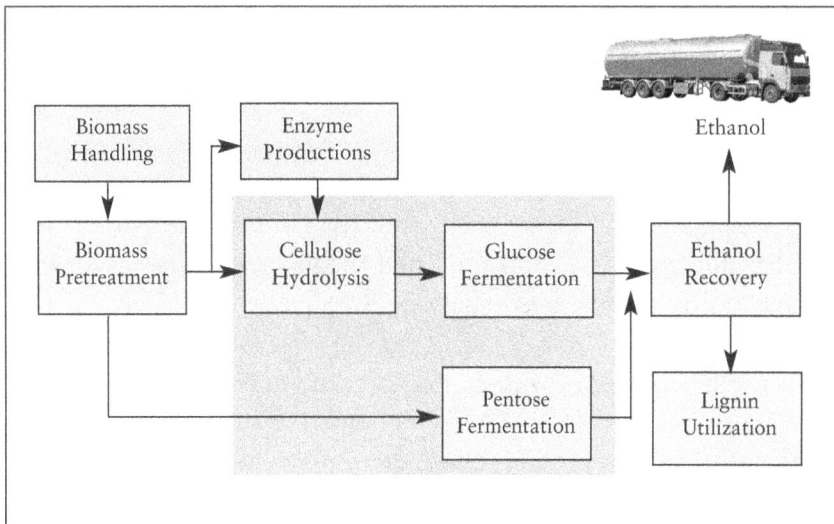

Figure 10.4 Producing from Cellulosic Biomass
Source: Renewable Fuels Association

identical to ethanol from other sources, such as corn starch or sugar, but has the advantage that the lignocellulose raw material is highly abundant and distributed diversely. It does require, however, a greater amount of processing to make the sugar monomers available to the microorganisms that are typically used to produce ethanol by fermentation.

Cellulosic ethanol is attractive because the feedstock, which includes wheat straw, corn stover, grass, and wood chips, is cheap and abundant. Converting it into ethanol requires less fossil fuel, so it can have a bigger effect than corn ethanol on reducing greenhouse-gas emissions. In addition, a hectare (ha) of grasses or other crops grown specifically to make ethanol could produce more than two times the number of gallons of ethanol as a hectare of corn. This is because the whole plant can be used instead of just the grain. Fast growing trees and woodchips are potential sources of cellulosic ethanol. Switchgrass is one of the main biomass materials being studied today because of its high levels of cellulose.[2] The nascent cellulose industry is seeing the installation of the operation of a number of new demonstration plants particularly in the United States. Relevant to countries like Jamaica which produce sugar cane, George Philippidis of Florida International University states, "Cellulosic ethanol from sugar cane bagasse represents one of the most promising options for our future energy needs. Therefore, understanding how to integrate cellulosic technologies into sugar mills is of keen interest to the world's investment community."[3]

Biomass Sources in Jamaica

There could be significant production of biomass for energy in Jamaica, which has biomass sources apart from agricultural by-products such as bagasse and agro industrial waste such as vinasse (dunder). Other sources include municipal and domestic waste, aquatic biomass (for example spirulina and water hyacinth) and wood and woody materials, including wood chips from the lumber industry.

Better use can be made of bagasse at sugar factories to improve electricity production and provide excess electricity to the utility grid system. Four main prospects exist with the optimistically estimated amount of electricity that could be made available to the utility: Frome (15 MW), Monymusk (10 MW), Bernard Lodge (5 MW), and Long Pond (5 MW). At Appleton, a private industry, an important positive environmental impact could arise from the anaerobic treatment of distillery wastewater in such sugar-based facilities as Appleton, Monymusk and Bernard Lodge (Rintala and Puhakko, 1994). In some cases, such as Appleton, this distillery waste (vinasse or dunder) is piped directly into sinkholes that lead to the river where the high biochemical oxygen demand (BOD) is deleterious to aquatic life. With treatment, an average of 90 per cent of the BOD could be removed. Clearly, waste water which can be subjected to anaerobic digestion and converted to energy not only solves an environmental waste disposal problem, but provides an economic benefit in the ancillary production of energy.

At present, wood for energy can compete only where production exceeds 10 dry tonnes/ha/year and the markets are local. In 1999, the trials at Font Hill demonstrated the good yields which can be achieved on such sites. Only species which coppice (can be cut and regrown) are able to give adequate yields and intensive management is required, including inputs of fertilizers and early prevention of weed competition. There are many advantages to coppice crops for energy, especially in that energy trees are sustainable, and energy crops have a high density (about 10,000 trees/ha) and short rotation cycle (3–5 years).

There is therefore a significant source of biomass in Jamaica, and biomass conversion systems have a role to play in reducing conventional energy demand and increasing energy supply in the Caribbean. This role will be heightened as national policies encourage greater reliance on renewable energy, with its benefits to society and the environment.

Environmental, Social and Economic Considerations

Bioenergy has significant environmental advantages. Biomass is often regarded as 'non-polluting' despite emitting large amounts of carbon dioxide and other pollutants, because the amount of carbon dioxide emitted in the combustion process equals the amount absorbed from the atmosphere during the growth of plants and the photosynthesis of carbon dioxide in the atmosphere. The burning of biomass produces less pollution than burning coal because biomass has a much lower content of sulphur and ash than coal. In addition, burning biomass does not increase the net concentration of carbon dioxide in the atmosphere because an equivalent amount of carbon dioxide is needed for plant growth. Successful breeding techniques can encourage a dramatic growth in trees. In addition, effective biotechnology programmes avoid the harmful ecological and health effects of pesticides.

Several developing bioenergy technologies also return much of the nutrients to the soil because they are captured in leftovers like ash from gasification and biogas slurry. Grasses like switchgrass can also be harvested regularly with little or no impact on the soil, particularly in the fall when the nutrients have returned to the roots, and if a portion of the stalks is left standing. Some grasses in fact develop such a rich root system that they actually contribute each year to increasing carbon and nutrients in the soil even while being harvested.

There are also major environmental and social challenges relating to biofuel production. Challenges relating to the environment are waste management, chemicals use, soil protection, water conservation, atmospheric emissions, use of genetically modified organisms (GMO) and biodiversity issues relating to the clearance of virgin forests. There may be an opportunity cost: the land used to produce biofuel could be used to grow food. In addition, there may be concerns related to working conditions. Because of these challenges, biofuels have gone from being perceived as a saviour to a significant environmental problem in just a few short years. There are clear examples of environmental degradation caused by biofuels, one example being the beleaguered peatlands of several countries in South Eastern Asia which have long been under threat from loggers seeking good financial returns from the high-value meranti and ramin trees that thrive in these precious habitats (Rowe, 2008).

There is also an obvious environmental danger in chopping down large numbers of trees for charcoal. Large quantities of wood are consumed each year in many developing countries, the wood is burned to produce charcoal and used in domestic cooking. The continued use of wood for charcoal at the present rate without replenishment will lead to the deforestation and de-beautification of countries, affecting not only food production and the environment but also tourism. In addition, the land clearing for biofuel production can cause in some cases considerable emissions ("biofuel carbon debt"), the compensation of which with biofuel use replacing fossil fuel can take long time spans (Searchinger et al., 2008).

Whether the impact of bioenergy on the environment is positive or negative depends on the technology and how it is utilized. If the entire world used fuel wood for cooking, for example, there could be a global ecological collapse. The world's forests would need to be razed to obtain enough energy. However, if countries gasified their industrial and urban wastes to heat their homes and generate their electricity, this would help to stabilize global atmospheric carbon levels and create sustainable societies. The difference in these scenarios is the increase in efficiency. Combined heat and power systems powered by biomass gasification utilize 70 per cent or more of the energy in the biomass, while campfires capture less than 10 per cent.

If biomass is environmentally sustainable, it is also economically viable. Using biomass that would otherwise rot or be incinerated is a form of recycling (Firn, 2008), and efficient projects also ensure the restoration of land to ensure future supplies of biomass. Efficent biofuel projects can also create many new jobs: biofuel projects are often labour intensive and can significantly increase employment in the agricultural sector.

Brazil shows how bioenergy is used for socio-economic and environmental benefits, as evidenced by the following quotation from a 2007 publication by Costa, Cohen and Schaeffer:

> In July 2003, the Ministry of Mines and Energy (MME) launched a biodiesel programme to gradually blend biodiesel, made from various indigenous vegetable species – mamona and palm oil, among others – with regular diesel fuel and increase its use in the transport sector. The principal objectives of the programme are environmental advantages and job creation. Among the products that can be obtained are B2, diesel with 2 per cent biodiesel content, B5, diesel with 5per cent biodiesel content, and so on up to B100.
>
> In December 2004, the federal government authorized the commercial use of B2. Using the 2002 Brazilian diesel consumption of 38 billion litres as a reference, the equivalent use of B2 would require the production of 800 thousand litres of biodiesel, which would reduce the diesel imports per year by US$160 million. (MME, 2004) According to the biodiesel programme, biodiesel producers that buy mamona and palm oil from small rural producers in the North and Northeast regions can apply for tax breaks.
>
> It is estimated that in the first phase an area of 1.5 million ha can be cultivated for B2 production. Each family producing 5 ha of mamona, with an average of 700–1,200 kg per ha (the production mamona is estimated in the range of 600 kg/ha to 1,200 kg/ha, without irrigation in the Northeast region of Brazil), can earn 2,500–3,500 Brazilian Reais (R$) just selling the seeds (primary production with no additional aggregated value), which gives a range of 160–275 Reais (R$) per month (MDA, 2004). The upper value is slightly higher than the national minimum wage per month as of December 2004.
>
> In the North region, palm oil is one of the best options for meeting biodiesel demand, with a potential of 3.2 million tonnes of production on 720,000 ha, sufficient for 140,000 families. The net income of a 5-ha family unit can reach six times the minimum wage. For soybean (Glycine max), a well-established agro-industrial

Palm oil has application as a source of biodiesel

sector in the country, the potential for B5 is around 1.8 billion litres of oil, requiring the cultivation of an additional 3 million ha (probably more area would be required), leading to 234,000 permanent jobs (Peres, 2004). However, as there are environmental and economic arguments against the uncontrolled expansion of soybeans in Amazonia, biodiesel production from soybeans should be compared with the cultivation of other species that perhaps create more jobs with less impact on the environment and certainly have a much larger oil production per ha. It should be noted that the production of biodiesel will require not only the oleaginous crops, but also sugarcane.

If Brazil is to be successful in its biodiesel programme, appropriate vegetable species need to be cultivated in different regions in the coming years. Rough estimates indicate that for a demand level of 1.85 billion litres of B5 produced from mamona, dendê and soybeans, around 1.0 to 1.2 million jobs can be created. The figures are much higher for B20; even in this case, the cultivated area, almost 20 million ha, is only about 25 per cent of the estimated area suitable for energy crops in the country. From the supply side, it is reasonable to point out that, considering the potential for mamona production in the Northeast region, some 2 million new jobs could be created. In summary, biodiesel production presents many environmental and social benefits and can be a unique alternative for increasing the number and quality of rural jobs in Brazil.

A large quantity of biomass matter may be available in agricultural communities following a harvest, such as after the processing of sugar cane (Ronco Consulting Corporation, 1986). Because biomass is seasonal, it will only provide a sustainable livelihood for farmers if projects are planned effectively enough to ensure a steady supply of the feedstock.

As a domestic source of energy, biofuels will contribute to reduced import cost for energy, energy security and avoidance of the volatility of the fossil fuel market particularly the oil market. Bioenergy can have significant economic potential in both developed and developing countries if a number of criteria for success are considered. The criteria are the energy contribution, potential risk and pay-off time of each project. The assessment, development, and utilization of biomass resources should coexist with the optimization of oil usage, and energy conservation to ensure a viable bioenergy operation.

The cost of electricity is a function of capital cost plus Operations and Maintenance (O&M) plus fuel cost, transport cost and biomass resource cost. Transport cost is dependent on the distance of the biomass resource from the power plant and the energy density of the biomass. The cost of the biomass is not simply the cost of production. For example, using corn for power generation would be a fuel cost similar to the market price for corn as a food crop rather than the actual cost of growing the crop.[4] On the other hand, using municipal solid waste has a negative fuel cost because burning it is a means of ultimate disposal. Biomass cogeneration plants seem to be the most attractive financially as long as there are no fuel costs and if the surplus electricity can be sold to the national grid.

Biomass cogeneration systems generate electricity and also utilize the waste heat from the electrical generation as industrial process heat. In the past, the main purpose of these biomass cogeneration systems was to produce process heat and perhaps generate some electricity to run the plant. Public utility organizations are assessing biomass cogenerating facilities as demand-side management systems and possible sources of electricity during periods of high demand. Public utility organizations should contract with their counterparts in private industry for the unused generating capacity of these systems. Private industry can upgrade the biomass cogenerating systems in order to produce electricity more efficiently.

Experience has shown that projects can be small and still be successful. Biomass electricity generating plants tend to be small because of the dispersed nature of the feedstock. Heat is generated at low pressure because the boilers cost far less than high-pressure units.[5] The boilers have low efficiencies in the range of 10 to 18 per cent. As a result, biomass plants have relied on low or zero cost of the biomass fuel for the operation to be economical. With the advent of biomass-integrated gasifiers/gas turbines, the unit cost of electricity production will decline in the future. Biomass is versatile and can be converted not only into electricity by burning but also to gaseous or liquid fuel by physical or biological means and it can often be stored. However, biomass has a lower energy density than fossil fuels. Hall et al. (1993) report heating values of 17.6 gigajoules per tonne for biomass (on a dry weight basis) compared to between 30 and 35 gigajoules per tonne for bituminous coals, and 23 to 26 gigajoules per tonne for lignite.

The economics of biomass generated electricity is also dependent on the source of the biomass, and in this vein, integrated waste-disposal/energy-generating projects are of special interest. Agricultural waste has significant economic potential and referred to as second-generation biofuels. The cheapest form of biomass is a crop residue such as sugar cane (bagasse), rice husks, and wood chips. Both heat and power can be generated for use within a factory or other building, and any excess electricity can be sold to the local power company. This is the cogeneration approach which can be highly cost-effective if the residue has no other value and if a reasonable price is obtained from the utility.

The average privately owned biomass facility will charge at least 7 US cents per kWh to be profitable. But calculating the cost of electricity may not always be straightforward or constant because calculating net energy efficiency involves many assumptions, not only about how crops are grown, harvested, and converted but also what resource use is avoided. Biomass harvesting is economical only if it does not diminish a soil's fertility or other properties influencing productivity. Biomass has significant economic and environmental value when used in the restoration of degraded land, as an adjunct to afforestation schemes on marginal lands and where the production of a crop is in excess of requirements. Decisions on the best use of land should be taken on a case to case basis after studying the alternative economics.

Investments in biomass generating plants have long-term payback periods but are necessary for achieving goals in the future, for example, by shifting from an old technology to a newer one (Edinger and Kaul, 2003). The strategy for biomass production and utilization should be congruent with the resources of the country. The overall strategy should include the introduction and adaptation of appropriate technologies for the production and consumption of firewood, charcoal, factory wastes and biogas.

It will also mean expediting the avenues of communication between forest and energy studies, as well as research related to biomass supply. Biomass electricity generating plants require the involvement of all stakeholders: producers, auto industry, fuel blenders, distributors and environmental agencies.[6]

Waste To Energy

Waste-to-Energy (WTE) is critically important because many landfills around the world are nearing capacity, thereby posing a serious environmental threat. Municipal waste is disposed of in valleys and specifically prepared sites and the bacteria decomposing the waste produce methane. Over 70 per cent of municipal solid waste consists of organic materials such as paper, food wastes, yard wastes, and plastic that have Btu combustion values. (Table 10.2)

Municipal waste is perhaps the second largest source of biomass power and is an energy source yet to be used in the Caribbean. WTE systems provide a disposal alternative. The best WTE systems start with a recycling programme which can become part of the waste sorting strategy.

The burning of landfill-generated methane is a potential fuel source in the Caribbean. Landfill gas is used in other countries for electricity generation and for firing brick kilns to produce steam in industry. There are about 250 landfill gas projects operating worldwide. The largest is in New York, USA, with production of around 150,000 cubic metres of gas a day, generating 80 MW of power.[7]

Table 10.2   Energy Value Index (million Btus per tonne)

Waste Element	Energy supplied if burned	Energy for virgin manufacture	Energy for recycle manufacture	Energy saved if recycled
Newsprint	8	27	22	5
Corrugated paper	7	17	17	0
Tissue Paper	8	12	14	-2
Aluminum	0	100	5	95
Steel	0	48	23	25
Glass	0	10	7	3

Waste-to-energy technology involves converting various elements of municipal solid waste such as paper, plastics, and woods to generate energy by either thermochemical or biochemical processes. The thermochemical techniques consist of combustion, gasification, and pyrolysis that produce high heat in fast reaction times. The biochemical processes consist of anaerobic digestion, hydrolysis, and fermentation using enzymes that produce low heat in slow reaction times.

The most common application of waste to-energy technology is combustion: the burning of municipal solid waste to produce steam for heating or to generate electricity. The combustion method captures heat energy by generating steam that can be used for space heating, and provides process heat for industrial operations or electricity generation. There are several types of combustion technology.

The options are:

- **Mass burn.** A mass burn waste combustor has a single combustion chamber with an on-site energy recovery mechanism. While an incinerator alone is not classified as a waste-to-energy technology, by attaching an additional heat recovery unit it can be considered as waste-to-energy technology.
- **Modular.** A modular waste combustor has a two- (or more) stage combustion unit and an energy-recovery unit. They are pre-fabricated and field erected for site construction.
- **Refuse-derived fuel.** A refuse-derived fuel system is an energy facility with extensive front-end processing used to pretreat waste. Such a system has a dedicated boiler for combusting prepared fuel.

Before considering any application of the waste-to-energy technologies, a comprehensive municipal solid waste management strategy must be developed. The economic feasibility of a waste-to-energy plant depends on the volumes of waste generated and its waste management costs. The waste-management cycle consists of collection, transportation, and disposal of the waste. The disposal method is pivotal since it influences how waste is collected and how far it must be transported. The costs of waste management can be substantial, in excess of millions of dollars per year for many Navy bases.

There are three waste disposal options: landfilling, converting the waste to usable material through recycling and converting waste into energy. A waste-to-energy plant is an excellent alternative to developing a solid waste disposal plant if the landfill option becomes too expensive.

A waste-to-energy plant can reduce the volume of waste by as much as 90 per cent. If there is a rapid increase in refuse disposal costs to a point at which it is no longer cost effective to continue off-site landfilling, waste-to-energy application should be considered.

To operate a waste-to-energy plant properly, there must be an effective waste management programme that includes recycling. Waste must be sorted and analysed for its Btu heat content. Its flow of volume must be sufficiently steady to meet the plant's design criteria before it is fed into the combustion chamber.

The financial attractiveness of a waste-to-energy facility hinges on many factors. Those factors include local landfill tipping fees, trash transportation costs, construction and operational costs, the purchase price of produced energy, recycling revenues and interest rates.

The City of Edmonton, Canada, serves as an important example of an efficient waste-to-energy system, having one of the best integrated waste management systems in the world, serving as an important illustration for the successful management of waste. Edmonton has all three components of integrated waste management (IWM), which are source reduction, reuse and recycling. The City redefined its IWM practices with significant developments in composting to divert residential waste from landfill sites.

Edmonton residents separate waste into recyclable and solid wastes. Recyclables are delivered to and handled by the material recovery facility (MRF) and solid wastes are delivered to and processed by composting facility. The MRF recovers newspaper, cardboards, mixed papers, milk cartons and jugs, glass, plastic, metals, and reuseable containers: the first three items account for more than 80 per cent of the recyclables. The composting facility is designed to handle between 180,000 and 200,000 tonnes of residential waste and 22,500 tonnes of biosolids. So far, the facility has processed between 126,000 and 167,000 metric tonnes of residential waste and up to 31,500 metric tonnes of biosolids. The facility has diverted about 55 per cent of household wastes annually.

The city also recycles concrete and asphalt by crushing them and using the material for the construction of the city's new roadway base and sidewalks. This was started in 1980 utilizing one large and one small recycling system. The larger system is operated by the Transportation and Street Department (TSD) and the smaller system is operated by the Waste Management Branch (WMB) at the Edmonton Waste Management Centre (EWMC). The TSD recycled about 180,000 tonnes of crushed material in 2006 and 2007. In addition, the city collects household hazardous waste (HHW). HHW is collected at its comprehensive HHW drop-off Eco Stations (ES). These facilities collect both HHW and inert materials, including flammables, oxidizers, corrosives, used oil and filters, paint, batteries, propane bottles and tanks, scrap metals, tire and other discarded materials, most of which are recyclable.

Ethanol

The biofuels industry is led by the use of ethanol in gasoline as a supplemental fuel and octane enhancer. As an octane enhancer, ethanol also known as ethyl alcohol replaces Methyl Butyl Ether (MTBE) which has been found to be environmentally unfriendly because it does not dissipate readily in ground water. Among the forces driving biofuel development are the shortage of oil refining capacity worldwide, the economic growth and fuel demand in China and India in particular and the rising demand for gasoline in the USA. Ethanol use is predicated on three factors: the sustainability of petroleum supplies, low incomes in the agricultural sector and environmental concerns relating to climate change. Jamaica has embarked on a strong biofuels programme with the introduction of 10 per cent ethanol in gasoline. Ethanol used in Jamaica comes from two main sources: wine alcohol imported from Europe and wet alcohol produced from sugar cane in Brazil. This wet alcohol is dehydrated to anhydrous alcohol (ethanol) in at least three facilities. The intention is to produce ethanol from local sugar cane feedstock. Ethanol has been introduced in a 10 per cent mix with gasoline (E10) in Jamaica, a fuel comprised of 85 per cent ethanol (E85) is a possibility over the longer term.

Ethanol is produced commercially from both corn and sugar cane. The energy input required to produce ethanol from corn is significantly more than that required to produce ethanol from sugar cane. It takes one unit of energy to produce three units of ethanol from corn. In comparison, one unit of energy produces an output of eight units of ethanol produced from sugar cane. Thus, sugar cane is a more efficient provider of ethanol.

Ethanol contains 35 per cent oxygen. The addition of oxygen to fuel results in more complete fuel combustion and reduces harmful tailpipe emissions. Gasoline containing a 10 per cent ethanol blend will reduce smog generating emissions such as carbon monoxide (25–30 per cent), particulate matter (40 per cent) and volatile organic components (7 per

Sugar cane is the most energy efficient source of the biofuel ethanol

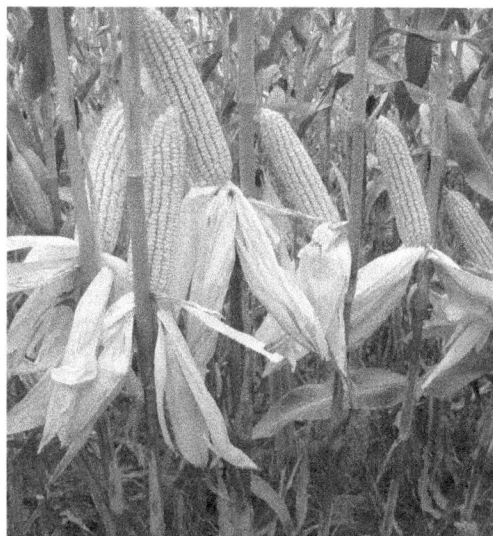

Corn is converted to ethanol but the conversion is less efficient than that of sugar cane. Also corn is a major food source so the use of corn introduces the conflict of food with fuel

cent). Ethanol can be used in place of harmful aromatics like benzene, the most toxic compound in some gasolines. In addition, ethanol does not contain sulphur.

At present, world ethanol production is only replacing 2 per cent of gasoline consumption, but global ethanol demand is growing by about 11 per cent per annum and about 50 billion litres of ethanol will be needed by 2010. If supply does not keep up with demand, ethanol prices will rise. Ethanol prices are likely to rise as demand grows particularly in China, North America and Europe. The increase in demand for ethanol has a direct effect on food prices. Using the corn industry in the USA as an example, one is seeing the price of food products that are corn based rising as the competition for corn supplies between food and ethanol increases.

The European Market has been proactive by providing African, Caribbean and Pacific (ACP) countries with duty free status for ethanol, an agreement which may change under the new Economic Partnership Agreement (EPA). Jamaica, the Dominican Republic, Barbados and other Caribbean countries, aware of the benefits of sugar cane ethanol, are already planning to restructure their sugar industry, inclusive of considering the possibilities of biomass cogeneration. Jamaica has privatized its sugar factories with a thrust towards ethanol production.

Brazil's Use of Ethanol

Previously in this chapter, the environmental and socioeconomic advantages of bioenergy in Brazil was discussed. Brazil has made significant efforts in the use of ethanol as a transport fuel with mixed economic results since the price of oil has decreased and the price of sugar has risen (Goldenberg et al., 1993). The country produces approximately 13 x 106 tonnes of ethanol per year. Brazil is known as an important pioneer in ethanol usage: 100 per cent ethanol is also used as a fuel. Flexi Cars using 100 per cent ethanol entered the Brazilian market in 2003. Presently, almost 100 per cent of all cars manufactured in Brazil are flexi cars capable of running on 100 per cent ethanol.

Ethanol has had a long history of use in Brazil. Ethanol has been used in the country since the end of 1920s and in 1931 the Government mandated 5 per cent blend of ethanol in all imported gasoline; in 1938 this mandate was extended to all gasoline consumed in the country. But it was after the launching of the National Alcohol Program (PROAL-COOL), that ethanol became an official and sustained motor fuel. The motivation of the government to establish this Programme was to reduce its massive reliance on oil imports that were posing a heavy burden on the country's balance of payments.

One of the most important lessons learned in assessing the Brazilian experiences with biofuels use is that specifications for each biofuel and its blends must be very well defined and detailed. The Brazilian anhydrous and hydrous fuel ethanol specifications are the result of a long experience with production and use of these fuels. Today it is widely accepted by the world auto industry that the cars can take up to 10 per cent by volume ethanol/gasoline blends without problems. Table 10.3 presents the Brazilian ethanol specifications.

It is important to note that there are environmental disadvantages to biofuels. In Brazil and other Latin American countries, there is the fear that biofuels may result in drastic deforestation and limit biodiversity by creating a monoculture. All ramifications should therefore be assessed before embarking on a biofuels industry.

Biodiesel

Biodiesel is a younger fuel than ethanol. Biodiesel is a non-petroleum-based diesel fuel consisting of short chain alkyl (methyl or ethyl) esters,

Table 10.3   Anhydrous and Hydrous Ethanol Specifications

Characteristic	Unit	Specification	
		Anhydrous ethanol	Hydrous ethanol
Acidity (as acetic acid)	mg/L	≤ 30.0	≤ 30.0
Electric conductivity	S/m	≤ 500	≤ 500
Specific Weight (200C)	kg/m3	≤ 791.5	809.3 ± 1.7
Ethanol Content	% v/v	≤ 99.3	93.2 ± 0.6
Hydrogen Potential (pH)	–	–	6.0 to 8.0
Evaporation Residue	mg/100mL	–	≤ 5.0
Ion Chloride	mg/kg	–	≤ 1.0
Ion Sulphate	mg/kg	–	≤ 4.0
Iron	mg/kg	–	≤ 5.0
Sodium	mg/kg	–	≤ 2.0
Copper	mg/kg	–	≤ 0.07

Source: Joseph Jr, 2007

typically made by transesterification of vegetable oils or animal fats, which can be used (alone, or blended with conventional petroleum derived diesel) in unmodified diesel-engine vehicles. "Biodiesel" is standardized as mono-alkyl esters and other non-diesel fuels of biological origin are not included. The increased use of biofuels will reduce smog and green house gas emissions. Many countries that have signed onto the Kyoto Protocol view biofuels as a tool to assist in reducing greenhouse gas emissions to their committed rate.

The EU and Brazilian experience with biodiesel is illustrative of biodiesel's great utility. Biodiesel production and use started in some EU countries in the mid 1980s but the volume produced only became important after 2002 when oil prices increased. Germany is the main producer with more than 40 per cent of the world production and has a considerable spare capacity installed indicating the trend to increase production. The EU specification is a good reference for countries considering the use of this biofuel, but adaptations may be needed since the EU experience is almost entirely concentrated on rapeseed as raw material.

In Brazil, the biodiesel specification was modified to allow the use of other feedstocks, without compromising the engine performance and durability. The automakers have approved the use of up to 5 per cent blend with mineral diesel and are conducting tests to verify the acceptability of higher levels of biodiesel in this blend.

A combination of economic, climatic and political factors led to the initiation of the Biodiesel National Program in Brazil. The goal of the Program was to eliminate the diesel imports, to reach the total independence of oil and liquid fuels. The government also tried to use the programme to strengthen agriculture based on small growers, especially in the semi-arid region of Brazil, maximizing the social benefits. The Law 11097 of January 13 of 2005 mandated the use of a 5 per cent blend, by volume, of biodiesel in all diesel oil consumed in the country, starting January, 2013. The law also mandated the use of 2 per cent biodiesel blend starting January, 2008 and permitted the voluntary use of 2 per cent biodiesel blend between between January 1, 2005 and December 31, 2007 and 5 per cent between January 1, 2008 and December 31, 2012. In this way, the introduction of biodiesel would be done smoothly and provide a clear picture to the investors.

To favour the small producer of the semi-arid region, the government gave a tax exemption to the biodiesel made from feedstock cultivated in this type of agriculture, representing the equivalent of US$0.11 per litre if produced in the North, North East or semi-arid region. The biodiesel produced using feedstock from large farms outside the semi-arid region is taxed in the same way as mineral diesel. The rules for the "Social Stamp" that provided the full tax exemption were clearly established.[8]

The search for a more effective biofuel should be a priority among energy companies. New regulations in Britain require that biofuels comprise 5 per cent of the transport fuel mix by 2010, and the EU has mandated that by 2020 all cars must run on 20 per cent biodiesel. Biodiesel reduces carbon dioxide emissions by nearly 80 per cent compared with petroleum diesel, according to the US Energy Department.

Jatropha and Algae: New Wonders for Producing Biofuel

Jatropha Curcas (Jatropha) is a poisonous scrub like plant that is widespread in tropical climates in Central America, the Caribbean, Africa, China, and India. The plant is highly drought resistant and grows rapidly.

According to an article in the *Times* (July 28, 2007), every hectare of Jatropha can produce 2.7 tonnes of oil and about 4 tonnes of biomass. Every 8,000 hectares of the plant can run a 1.5 megawatt station, enough to power 2,500 homes.[9]

Jatropha was brought to Europe from Central America by Portuguese explorers in the sixteenth century and has since spread worldwide. Until recently, jatropha had few uses: malaria treatment, a windbreak for animals, live fencing and candle-making. An ingredient in folk remedies around the world, it earned the nickname "physic nut", but its sap is a skin irritant, and ingesting three untreated seeds can poison a person.

Conversely, in India about 11 million hectares have been identified as potential land on which to grow jatropha. The first jatropha-fuelled power station is expected to begin supplying electricity in Swaziland in 2010. Meanwhile, companies from Europe and India have begun buying up land in Africa as potential jatropha plantations. Nonetheless some other countries are cautious about Jatropha for other reasons.

Jatropha starts yielding from the second year onwards and produces for approximately 40 years. The oil extracted from the seeds requires little processing and modification before it can be used in diesel engines. When the seeds are crushed, the resulting jatropha oil can be burnt in a standard diesel car, while the residue can be processed into biomass to power electricity plants.

Other specifics regarding the production of Jatropha are summarized below:

- Jatropha needs at least 600 mm (23 in) of rain a year to thrive. However, it can survive three consecutive years of drought by dropping its leaves. It is excellent at preventing soil erosion, and the leaves that it drops act as soil-enriching mulch.
- The cost of 1,000 jatropha saplings (enough for one acre) in Pakistan is about £50, or 5p each. The cost of 1 kg of jatropha seeds in India is the equivalent of about 7p. Each jatropha seedling should be given an area two metres square.
- Twenty per cent of seedlings planted will not survive. Jatropha seedlings yield seeds in the first year after plantation.

Algae are among the fastest growing plants in the world, and about 50 per cent of their weight is oil. This lipid oil can be used to make biodiesel for land vehicles and airplanes (CNN, 2008). Algal energy does not require the utilization of agricultural land and water, and delivers 10 to 100 times more energy per acre than crop-based biofuels (Clean Technica, 2008). Algal oil has the potential to become an important ingredient in biofuel and could be somewhat cheaper if developed from local sources. Algae blended with ethanol can be used to produce biodiesel. However, studies are still at an experimental stage and there are presently no commercial applications.

The Routes to Biodiesel Production from Oils to Fats

Apart from Jatropha and algae, biodiesel can be produced from plants such as palm oil, castor beans, soya and rapeseed. There are three basic routes to biodiesel production from oils and fats:

- base catalyzed transesterification of the oil
- direct acid catalyzed transesterification of the oil
- conversion of the oil to its fatty acids and then to biodiesel

Almost all biodiesel is produced using base catalyzed transesterification as it is the most economical process requiring only low temperatures and pressures and producing a 98 per cent conversion yield.

The transesterification process is the reaction of a triglyceride (fat/oil) with an alcohol to form esters and glycerol. A triglyceride has a glycerine molecule as its base with three long chain fatty acids attached. The characteristics of the fat are determined by the nature of the fatty acids attached to the glycerine. The nature of the fatty acids can in turn affect the characteristics of the biodiesel. During the esterification process, the triglyceride is reacted with alcohol in the presence of a catalyst, usually a strong alkaline like sodium hydroxide. The alcohol reacts with the fatty acids to form the mono-alkyl ester, or biodiesel and crude glycerol. In most production methanol or ethanol is the alcohol used (methanol produces methyl esters, ethanol produces ethyl esters) and is base catalyzed by either potassium or sodium hydroxide. Potassium hydroxide has been found to be more suitable for the ethyl ester biodiesel production; either base can be used for the methyl ester. A common product of the transesterification process is Rape Methyl Ester (RME) produced from raw rapeseed oil reacted with methanol. The products of the reaction are the biodiesel itself and glycerol.

A successful transesterification reaction is signified by the separation of the ester and glycerol layers after the reaction time. The heavier, co-product, glycerol settles out and may be sold as it is or it may be purified for use in other industries, examples being the pharmaceutical and cosmetics industries.

Straight vegetable oil (SVO) can be used directly as a fossil diesel substitute, however using this fuel can lead to some engine problems. Due to its relatively high viscosity SVO leads to poor atomization of the fuel, incomplete combustion, choking of the fuel injectors, ring carbonization, and accumulation of fuel in the lubricating oil. The best method for solving these problems is the transesterification of the oil.

The engine combustion benefits of the transesterification of the oil are:

- lowered viscosity
- complete removal of the glycerides
- lowered boiling point
- lowered flash point
- lowered pour point

The Future of Bioenergy

The present situation of course guides the future of bioenergy. The present global situation, according to Moreira (2008) is that biomass currently provides around 46 EJ of bioenergy in the form of combustible biomass and wastes, liquid biofuels, renewable municipal solid waste (MSW), solid biomass/charcoal, and gaseous fuels. Biomass is the third most important new and renewable source of electricity at the global level and the second most important source in many developing countries. Biofuels in particular have become subject of much interest in the global energy sector because they are the only widespread commercially available alternative to fossil fuels. Globally, biomass-fueled heating still provides five times more heat than solar and geothermal combined, and continues to grow in northern Europe. Currently consumption of bioenergy is approximately five times greater than the energy provided by hydropower.

Several countries have set up regulatory frameworks for bioenergy, and are also providing incentives to support nascent biofuel industries. These developments are expected to spur a sustained worldwide demand and supply of bioenergy in the years to come. There has been significant growth in the production of biodiesel in a number of countries other than Brazil – including Argentina, Colombia, the USA and Germany. Because of this growth, it is very likely there will be global certification standards in the future.

There are many promising indications. For example, cellulosic ethanol could become a mainline biofuel source after 2020. A number of biorefineries have now been commissioned in the United States, funded in part by the Department of Energy to produce cellulosic ethanol commercially albeit on a small scale. The worldwide production of cellulosic ethanol will amount to at least 16.5 billion gallons in 2020, if the targets set in the United States, China, Europe, Japan and Brazil are achieved. Based on currently proposed and signed legislation, the United States would account for over 63.9 per cent of that market, while the EU and China would account for 10.4 and 11.5 per cent respectively.[16]

Green diesel also has promise: it is a new technology which has not yet been commercialized. Green diesel should not be confused with biodiesel. Green diesel is diesel oil manufactured using vegetable oils or fat oil directly through hydrogenation of the feedstock. The advantage of green diesel is the integration of the oil industry with biomass producers earlier in the process when compared with bioethanol. A greater number of companies will also become involved in technological innovations such as cold starch fermentation, corn fractionation and corn oil extraction. Companies are also utilizing biomass gasification and methane digesters to reduce natural gas consumption.

The benefits of bioenergy could be realized over the next few decades if sufficient research is devoted to promoting the development of the related technology; such as direct combustion generation of electricity and gasification. The current costs of these technologies would need to be lowered so that markets for biomass energy could expand. Indeed, the versatility of bioenergy projects regarding ecosystems, social adaptability and markets makes it a valuable alternative for the future of energy.

References

Ahmed, K. (1994) "Renewable Energy Technologies". World Bank Technical Paper no. 240.

Barnes, D.F., and Floor, W. (1999) "Biomass Energy and the Poor in the Developing World". *Journal of International Affairs* 53 (1) pp. 237–262.

Clean Technica (2008) "Algal Fuel One Step Closer To Becoming A Conventional Oil Alternative". Retrieved from: http://cleantechnica.com/2008/07/31/algal-fuel-one-step-closer-to-becoming-a-conventional-oil-alternative/

CNN (2008) Algae: "The Ultimate in Renewable Energy". Retrieved from: http://edition.cnn.com/2008/TECH/science/04/01/algae.oil/index.html

Edinger, R., and Kaul, S. (2003) *Sustainable Mobility: Renewable Energies for Powering Fuel Cell Vehicles:* Westport, Connecticut: Praeger.

Firn, R.D. (2008) "More on Biofuels". *World Watch* 21 (3) pp. 2–9.

Goldenberg, J., Monaco, L.C., and Macedo, I.C. (1993) *The Brazilian Fuel-Alcohol Program.* Washington D.C.: Island Press.

Hall, D.O., Rosillo-Calle, F., Williams, R.H. and Woods, J. (1993) 'Biomass for Energy: Supply Prospects'. In T.B. Johansson and others, eds. *Renewable Energy Sources for Fuels and Electricity.* Washington D.C.: Island Press.

Joseph, H. (2007) The Vehicle Adaptation to Ethanol Fuel. London: The Royal Society International Biofuels Opportunities Event.

Luo, Y. (1994) "The Engineering and Economic Evaluation for Chemical Utilization of Biomass". *Renewable Energy* 5 (pt. II) pp. 866–874.

Macedo da Costa, M., Cohen, C., and Schaeffer, R. (2007) "Social Features of Energy Production and Use in Brazil: Goals for a Sustainable Energy Future".

Natural Resources Forum 31 (2007) pp. 11–20.

Moreira J. R. (2008) "Biomass for energy: Uses, Present Market, Potential and Costs". National Reference Center on Biomass – CENBIO, Institute of Electrotechnology and Energy.

Renewable Fuels Association. Retrieved from: http://www.ethanolrfa.org/resource/made/

Rintala, J.A., and Puhakko, J.A. (1994) "Anaerobic Treatment in Pulp and Paper-Mill Waste Management: a View". *Bioresource Technology* 47, pp. 1–18.

Ronco Consulting Corporation (1986) "Electrical Power from Cane Residues in Thailand – a Technical and Economic Analysis". USAID Report No. 89–03.

Rowe, M. (2008) "Fuelling the Debate". *Geographical* 80 (2) pp. 44–54.

Sayigh, A. (1995) "Prospects and Achievements of Renewable Energy".

Proceedings of the Caribbean High Level Workshop on Renewable Energy Technologies St UNESCO, pp. 37–55.

Searchinger, T., Heimlich, R., Houghton, R.A., Dong, F., Elobeid, A., Fabiosa, J., Tokgoz, S., Hayes, D., and Yu, T.H. (2008) Use of U.S. Croplands for Biofuels – Increases Greenhouse Gases through Emissions from land-use Change. *Science* 319, pp.1238–1240.

World Bank (1994) Making Development Sustainable. Washington D.C.: World Bank.

11. | Hydropower

A Low Cost Renewable Resource

WATER CONSTANTLY MOVES THROUGH A vast cycle. It evaporates from oceans and inland water bodies, forms clouds, precipitates as rain and then flows back to the sea. The energy of this water cycle, which is driven by the sun, is tapped most efficiently with hydropower. From an environmental viewpoint, hydropower is clean power. It produces no carbon dioxide, sulphur dioxide, nitrous oxides or other air emissions. Hydropower plants also produce no liquid or solid wastes. The major problem is that of resource allocation. Water has many uses and benefits, so there may be conflicting demands on the resources available.

According to the Lafitte (n.d.), the world hydro potential is significant, as indicated by the numbers below:

- Gross hydro potential: 40,500 TWh/year
- Technically feasible: 14,300 TWh/year
- Economically feasible: 8,100 TWh/year

The installed capacity is approximately 700 GW (corresponding to the 2,600 TWh/year) and the remaining exploitable capacity is 1,500 GW (producing 5,500 TWh/year). By the middle of the twenty-first century, world energy consumption could be multiplied by a factor of 2.5 to 3.0, largely attributable to hydropower.

Hydropower has been used since antiquity. The first water wheel was recorded in 86 BC. Subsequently, there was the 'Greek' mill and later the 'Roman' mill, a water wheel with the addition of a gear. During the eighteenth century, the word turbine appears in the literature and these machines later became the basis of electricity generation from water.

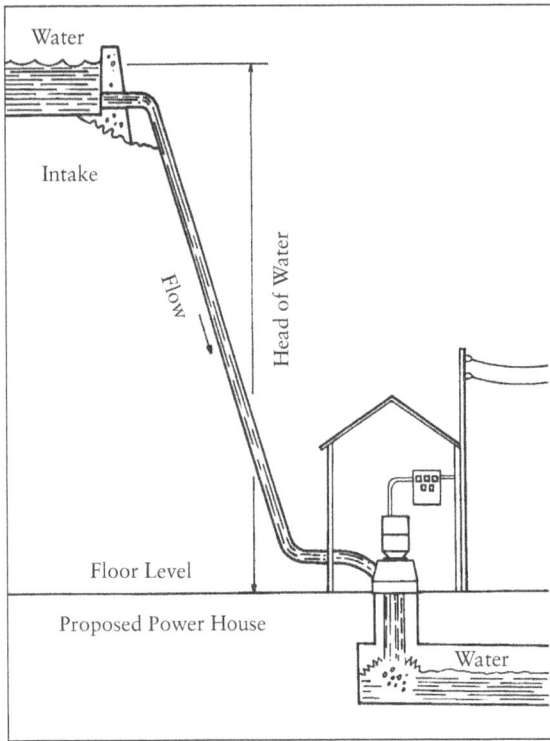

Figure 11.1 Simple Hydropower Configuration

Turbines are a specialized type of engine, like those used in jet airplanes, and are excellent prime movers.

Although the water turbine is the simplest form of prime mover, the generation of power by this means is little understood compared with internal combustion and reciprocating steam engines. A simple and basic principle is that more power cannot be obtained from the shaft than is put into the inlet (Figure 11.1). The power available depends on the 'flow' and head of water at the site. Flow varies with rainfall and is measured in litres per second or cubic feet per second. Head is the vertical distance from the water level at the intake to the turbine position. This can range from 5 to more than 100 metres. Sites with heights less than 5 m are regarded as 'low head' and are likely to produce small amounts of electrical output (Duckers, 1995).

Important civil works have to be built at a hydropower site. There are intake structures, diversion dams, canals and penstocks, the latter carry-

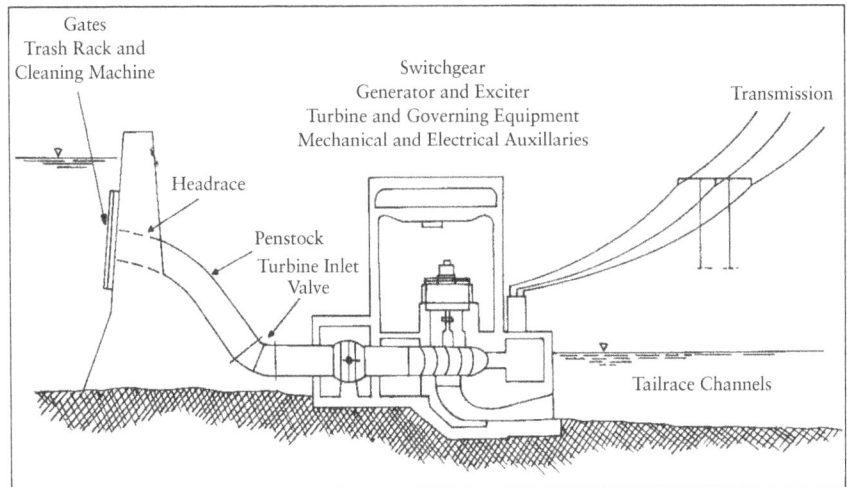

Figure 11.2 Schematics of a Hydropower Plant

ing water to the turbines. The tailrace returns the water to the river. There are also the electrical systems including generator, controls and switchgear, as illustrated in Figure 11.2.

Types of Power Plants

Conventional hydropower plants are of three types: storage, run-of-river and diversion. The geographic and hydrologic characteristics of particular sites dictate the appropriate type of hydroelectric development. Storage plants have reservoirs containing water from the incoming stream flow (Figure 11.2). Such plants are usually multipurpose facilities, built for flood control, irrigation, water supply, perhaps recreation, in addition to electricity supply. Storage facilities make good baseload and peaking plants. They have a high capital cost but low operating costs make them more cost-effective when used continuously. They are also useful for peaking power because they can be stopped and started easily and at little cost. Because of their flexibility in meeting both peak and base load demand, storage systems are attractive for utility companies.

Surface water-flow data measurement is crucial to hydropower studies. Among the hydrologic information to be analysed is the flow duration curve, head versus flow curve, and wet year versus dry year flow. It is also necessary to know how rapidly surface water can run off during periods of flooding.

Run-of-river plants use natural streamflow for power generation, although a small dam is often used to augment the head in such projects. The rate at which water flows into the dam roughly equals the rate at which water flows through the plant. Therefore, such systems essentially impound no water. In some projects, only part of the flow is diverted to the turbines. Some plants use headponds to impound water behind the dam, thus storing enough energy to shift maximum power output to service peak electricity demand times. Run-of-river projects are generally smaller than reservoir storage plants. They are usually operated at baseload capacity, running continuously when sufficient water is available. In dry seasons, such units can serve as peaking capacity if they employ some pondage.

Diversion plants involve a man-made channel, or aqueduct, of sufficient slope to make enough head to drive the turbine. Most of these structures are built solely for hydroelectric power, although many diversion projects are sited at existing irrigation or water supply conduits. Typical

Dams are used to augment the hydropower potential of rivers. However dams may in certain physical circumstances have adverse social consequences in respect of land use and the required relocation of people

diversion projects have no storage capacity, although some have reservoirs, which provide storage capacity.

Large-scale schemes produce hundreds of MW. Small-scale schemes typically deliver less than 10 MW and micro schemes less than 100 kW (Table 11.1).

Environmentalists have not always viewed large-scale schemes with favour. Large dam projects, like China's Three Gorges, have flooded natural habitats to create reservoirs, and downstream ecosystems are permanently affected when rivers flow according to the plans of engineers rather than natural cycles. Because moving water is a powerful force, a growing number of homeowners are recognizing that they can use its natural energy to reduce or eliminate gas-belching power plants. The small-scale hydro schemes, referred to as microhydro may be an inexpensive substitute for homeowners who prefer to generate their own electricity (Vartain, 2008).

Table 11.1 Classification of Hydropower Schemes with Typical rather than Definitive Values

	Output	Head
Large Scale	Over 10 MW	50–1,000 m
Small Scale	Less than 10 MW	10–1,000 m
Micro	10–100 kW	5–30 m
Low head	10–100 kW	1–5 m

Source: Duckers, 1995

Turbines

Various types of turbines are used to produce hydraulic power, including the Pelton, Kaplan, and Francis designs (Figure 11.3). The simplest turbine is the Pelton wheel, which is found in small, high head plants (Figure 11.4). Water jets strike the concave blades (buckets) of the turning wheel in sequence, rotating the attached shaft in the generator. This is an impulse turbine and efficiency is high over a variety of flows.

There are more complicated turbines. A Kaplan turbine resembles a ship's propeller. Water from the penstock strikes the blades, turning the turbine shaft. A Kaplan machine is particularly useful for low head applications (less than 20 metres net head). The Francis turbine looks like a fallen-over water wheel. In this case, water from the penstock completely surrounds the turbine and provides a constant pressure around the wheel. Wicket wheels, which may have fixed or adjustable openings, control the

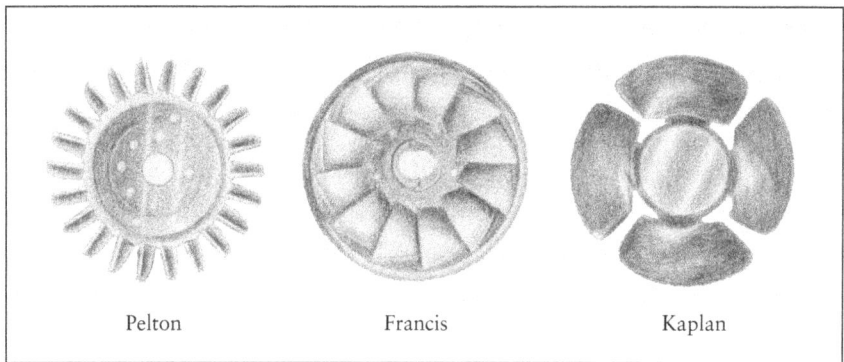

Pelton   Francis   Kaplan

Figure 11.3 Pelton, Francis and Kaplan Turbine Designs

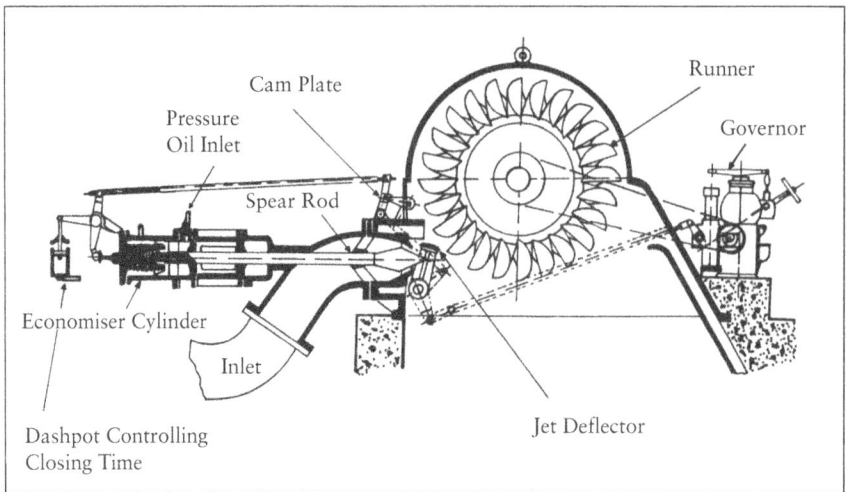

Figure 11.4 The Pelton Wheel

water flow. Water strikes the turbine blades, to which it transfers its energy, and exits through the middle of the turbine. The design is suitable for both low and high heads.

Costs

Table 11.2 gives a general idea of the percentage to costs of the constituent parts of a hydropower scheme for a 1 MW scheme. This scheme would cost about US$1.8 million, but local conditions may change the

Table 11.2   Relative Cost of a Typical 1 MW Hydropower Project

Feature	Percentage of Total Project Features		
	Low	Average	High
Headworks	4	15	8
Penstock	5	30	20
Powerhouse	3	20	9
Turbine Generator	24	77	47
Transmission	3	42	16
Total			100

Source: Duckers, 1995

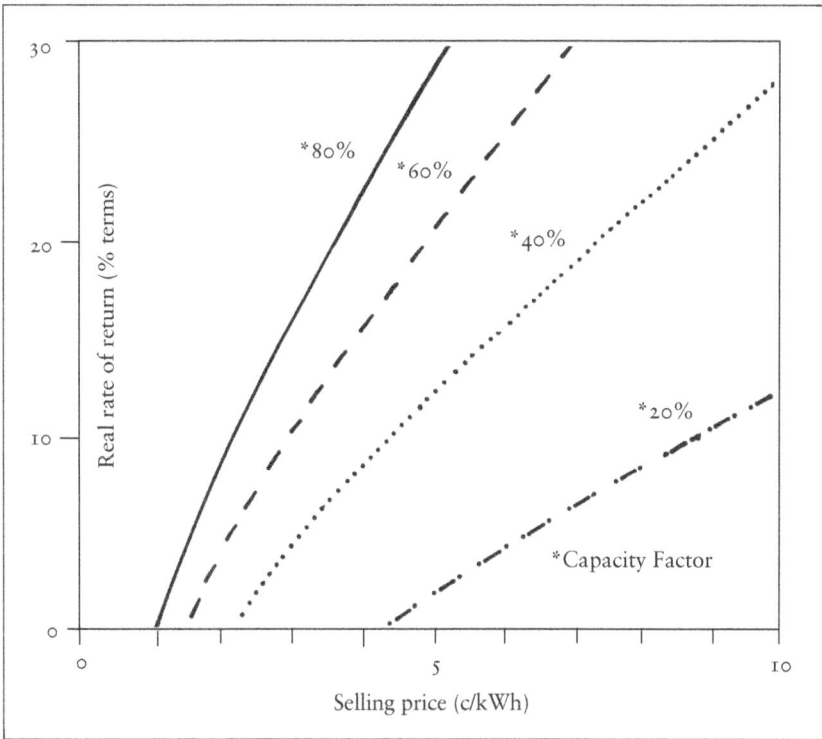

Figure 11.5 The Rate of Return from a Hydroelectric Project based on an Installed Cost of US$1,000/kW.

Source: Duckers, 1995

costing. Figure 11.5 gives the rate of return from a hydroelectric project based on an installed cost of US$1,000/kW.

The overall capital cost of hydropower plants tend to be greater than that of fossil fuel plants of comparable size. This is because the construction costs of the hydropower plant are normally higher. Installation costs can vary between US$1.5 and 4 million/MW. Civil engineering and other structural elements of a hydropower scheme tend to take up approximately 70 per cent of the installation cost. Operational costs are usually in the range 1–2 per cent of installation cost. Such high initial investment costs are usually a barrier despite low operating costs.

Hydropower plants have a long lifetime with low operating and maintenance costs. The operation and maintenance of a hydropower plant increase over time at an average rate of only one-third to one-half of the inflation rate. On a lifetime basis, hydropower projects are cost-competitive because there is no need to buy fuel from an outside source at a cost which is likely to escalate with inflation. Whereas a hydropower scheme can last for about 100 years, the upgrade and modifications to equipment would be expected to take place every 30–35 years.

The cost of energy may be represented by:

$$\frac{C \times R + O}{E \quad E}$$

where C = Capital cost

R = Capital recovery factor including investor return requirement, loan interest rates and amortization period.

E = Energy yield

O= Operation and maintenance cost.

When the often low external (environmental) costs are considered and if incentives such as premium power purchase rates are given, then hydropower can be an attractive energy option. But there are also social costs including a conflict in the use of the water resource.

The monetary and social costs of large-scale hydropower schemes suggest that a careful cost benefit evaluation should be performed. The primary purpose of energy projects is not the promotion of any given technology, but rather to meet the energy needs of the country in the most effective manner. When hydropower satisfies this criterion, it should obviously be used, as it is an important climate change mitigation technology.

References

Duckers, L. (1995) "Hydroelectricity. Proceedings of the Caribbean High Level Workshop on Renewable Energy Technologies" December 5–9, 1994, Saint Lucia. UNESCO, pp. 90–102.

Lafitte, R. (n.d.) "World Hydro Power Potential". International Sustainable Energy Organization (ISEO). Retrieved from: http://www.uniseo.org/hydropower.html

Vartain, S. (2008) "Wet and Wild: Small-Scale 'Microhydro' for the Home". *The Environmental Magazine*, May-June, 2007. Retrieved from: http://findarticles.com/p/articles/mi_m1594/is_3_18/ai_n27280486/

12. | Ocean Energy

An Emerging Energy Source

Ocean Energy represents one of the largest renewable resources available on the planet. Ocean Energy is an emerging industry that has the potential to satisfy worldwide demand for electricity, water and fuels, when coupled with secondary energy conversion principles.

Ocean Energy represents a number of energy conversion technologies, described below:

- Ocean Thermal Energy Conversion uses the temperature differential between cold water from the deep sea and warm surface water.
- Wave Energy is represented by surface and subsurface motion of the waves.
- Tidal Energy harvests the energy of ocean currents and tides.
- Osmotic Energy is the pressure differential between salt and fresh water.

Ocean Thermal Energy Conversion

Ocean Thermal Energy Conversion (OTEC) is a relatively new technology which is currently being tested for commercial use as a renewable energy source. The ultimate worldwide potential is approximately 810,000 TWh/yr. The OTEC energy generation system is dependent on the temperature difference that exists between warm surface waters and cold deep ocean waters. A temperature difference of at least 20°C is a prerequisite for a viable OTEC operation. This requirement is usually more available in tropical and sub-tropical seas.

In the tropics there is a significant temperature difference between surface and deep water, the latter being rich in nutrients such as nitrates and phosphates. If the warm and cold water (20°C difference) are brought together, a turbine can be operated to generate electricity by OTEC. There are two main methods of generating such energy, both of which utilize cold deep water. The Closed Cycle system uses a working fluid, such as ammonia or freon, which is evaporated in a heat exchanger through which warm water passes. Vapour is produced by pressure and is used to drive a turbine which generates electricity. The exhaust vapour from the turbine is condensed back into liquid in a heat exchanger through which cold water passes (Figures 12.1, 12.2).

Open Cycle OTEC uses sea water as the working fluid, the sea water being flash evaporated under a partial vacuum. The low pressure steam is passed through an extremely large turbine which produces energy.

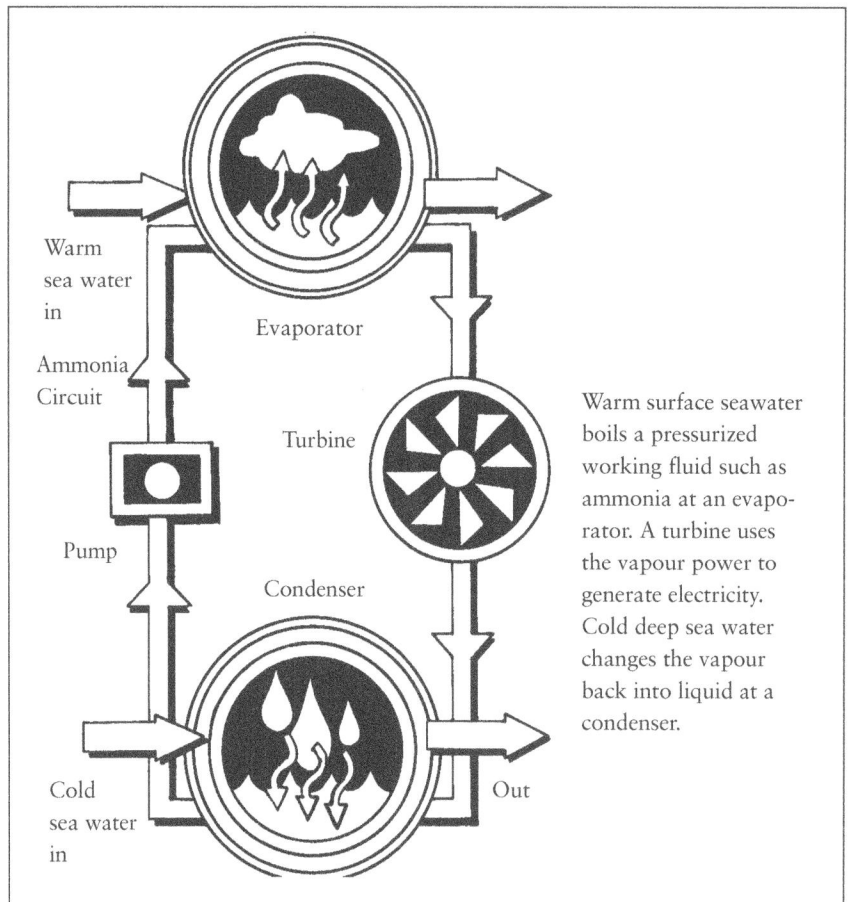

Figure 12.1 Principles of Closed-Cycle OTEC

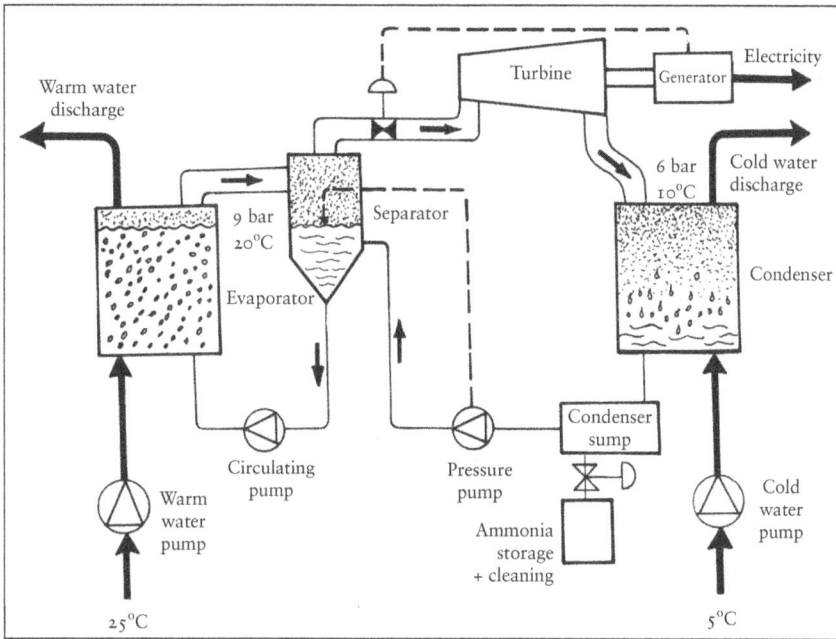

Figure 12.2 Schematics of Closed-Cycle OTEC

Spent vapour is cooled in a condenser and because the condensate is not returned directly to an evaporator the system is called open-cycle (Figure 12.3). Powerful vacuum pumps not only maintain the vacuum required to make the warm sea-water flash evaporate but also remove much of the large quantities of gases dissolved in the sea water. Desalinized water is a by-product of closed-cycle OTEC whereas aquaculture, air-conditioning and natural beta-carotene (for pharmaceutical use) are ancillary products of both closed-cycle and open-cycle OTEC. OTEC plants may be land based, shelf mounted or floating such as on a barge.

A Hybrid System is made possible by combining the features of both the closed-cycle and open-cycle systems. In a hybrid system, warm sea water enters a vacuum chamber where it is flash evaporated into steam, similar to the open-cycle evaporation process. The steam vaporizes a low boiling point fluid (in a closed-cycle loop) that drives a turbine to produce electricity.

Cost Effectiveness

The combined production of energy, fresh water and aquaculture from OTEC will ultimately become cost effective. Although OTEC has been

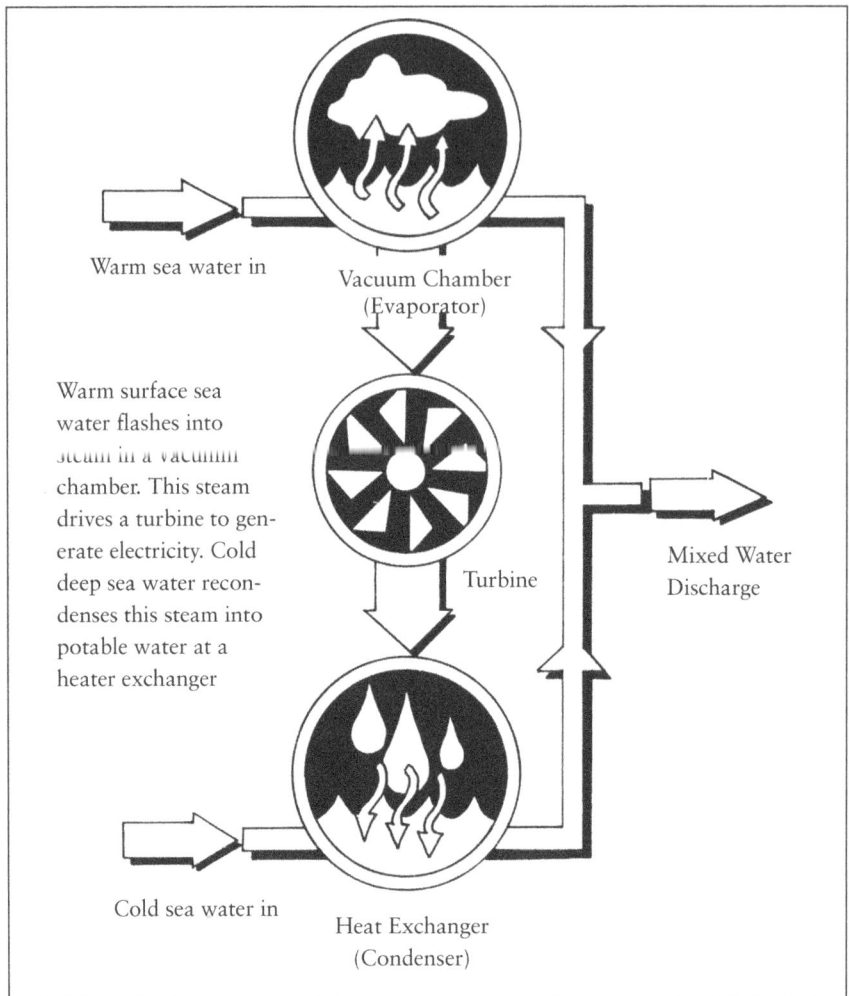

Warm sea water in

Vacuum Chamber
(Evaporator)

Warm surface sea
water flashes into
steam in a vacuum
chamber. This steam
drives a turbine to gen-
erate electricity. Cold
deep sea water recon-
denses this steam into
potable water at a
heater exchanger

Turbine

Mixed Water
Discharge

Cold sea water in

Heat Exchanger
(Condenser)

Figure 12.3 Principles of Open-Cycle OTEC

demonstrated experimentally to be a sound technology for producing positive net electric power, certain design and economic uncertainties have hindered its commercial development. The problem of financing huge OTEC power installations is also complicated by the fact that lending institutions first wish to be convinced that the technology is well-proven and that the operation of any OTEC plant is economically viable. Evidence has accumulated in the last five years to show that the economic viability of an OTEC plant can be significantly improved if, in addition to producing electricity, the by-products can also be profitably utilized. Many developing countries should proceed with a small pilot plant on a commercial scale which would provide sufficient operational information to plan, design, and finance large-scale commercial plants. The size of this early commercial OTEC pilot plant would be between 2 (megawatts) MW and 10 MW.

Electricity Generation

A key component of an OTEC system is a continuous supply of cold sea water pumped up from ocean depths. These deep ocean waters not only have low temperatures but are also rich in nutrients and are more or less sterile. The installation of a pipeline to pump up the cold water represents a significant technical challenge and, depending upon the coastal underwater topography, the length and diameter of piping required, flow rates and pumping power needed, it can become the most expensive component of any OTEC system. There are also challenges related to laying the pipeline which should be made from polyethylene and be relatively light. In addition, there are high costs in servicing and maintaining the pipeline over time.

The greatest advance in OTEC technology in recent years has been with the heat exchangers. Whereas costly titanium was used in the past, aluminium can now be used instead without any significant corrosion or bio-fouling problems. Titanium is more than ten times as costly as aluminium. Therefore, important cost savings can be realized without

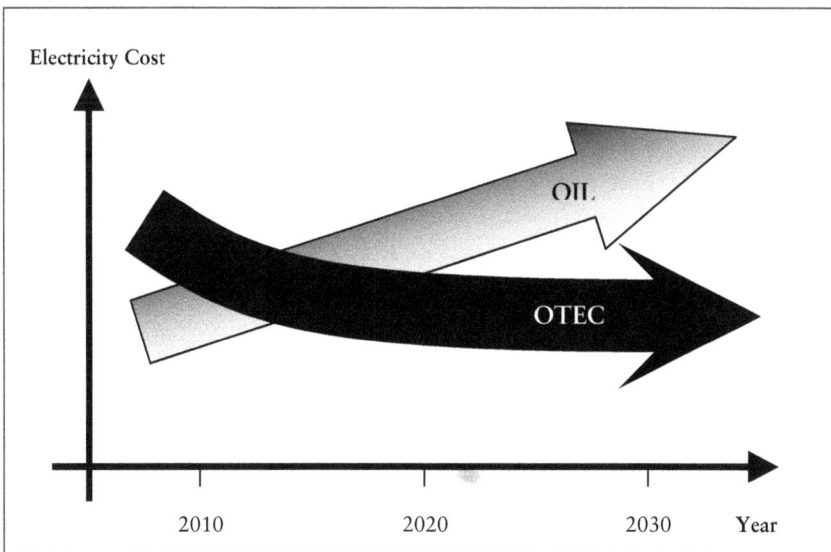

Figure 12.4 Gross Future Trends in Oil versus OTEC Prices

reducing the performance of the heat exchanger. ALCAN has demonstrated that significant reductions in cost can be achieved by the use of roll-bond heat exchangers which are relatively cheap and flexible to manufacture in different patterns and sizes. This cost is about four times lower than that of aluminium shell, tube and fin systems and more than twenty times less than using titanium. There have been few corrosion problems using the aluminium heat exchangers and the roll-bond system. Bio-fouling has been controlled by using chlorine (70ppbl1 hour/day) and is at acceptable levels. One result of this substitution is that a new 30 MW OTEC plant should be able to produce energy at a cost approximately equivalent to oil at US$50 per barrel. Figure 12.4 on page 139 shows possible trends in the price of oil versus OTEC.

Fresh Water Production

The early development of OTEC is based on a dual purpose role, namely the production of energy and fresh water. The high throughput rate (4,503 m per second for a 100 MW open-cycle plant) means that fresh water can be generated at rates sufficient to significantly assist municipal requirements (Tanner, 1995). In areas where fresh water and electricity are at a premium, for example, some Caribbean and Pacific Islands, OTEC generated fresh water is cheaper than water produced by other means, such as reverse osmosis.

A good example is provided from the Mediterranean, another region where OTEC is feasible. The Mediterranean Action Plan of the United Nations Environment Program (UNEP) indicates that water shortages in the region are likely to worsen as the old aquifers are drained. Malta is a classic case where new desalination technology is needed. Currently, the greater part of Malta's fresh water is desalinated at five plants which consume 15 per cent of the country's electricity production, thus making it the most expensive water in the region (Tanner, 1995).

A 40 MW hybrid plant located in Jamaica, for example at offshore Cow Bay in St Thomas, could provide 10 million gallons of water per day for the Kingston municipal water supply system. A number of other islands in the Caribbean, Pacific and Mediterranean have the bathymetric conditions suitable for OTEC technology. Japan has had an OTEC programme since 1974 in which Saga University has played a leading role. Initial studies focussed on a 100 MW plant but the more significant research was on mini-OTEC with closed-cycle systems on Nauru in the Pacific, Shimane in the Japan Sea and at Tokunoshima on the island of Kyushu. In recent years ongoing Japanese research and development programmes have been directed to design methods, advanced materials, and procedures.

Aquaculture

One of the most important components of an OTEC system is a continuous supply of cold sea water pumped up from ocean depths. These ocean waters not only have low temperatures but are also rich in nutrients. Compared to warm surface water, inorganic nitrate-nitrite values in deep cold water are 190 times higher, phosphate values 15 times higher and silicate values 25 times higher. Aquaculturists have long viewed such waters as a valuable resource which can be utilized for growing a mix of aquatic animals and plants. The ability of the OTEC system to provide flexible, accurate and consistent temperature control, high volume flow rates, and sea water that is relatively free of biological and chemical contaminants, can be translated into a saleable aquaculture resource.

Organisms such as salmon, trout, abalone, oyster, lobster, giant sea clam, sea urchin, seaweed and microalgae have been successfully grown in the OTEC cold water system. Nonetheless, the commercial viability of some of these aquaculture operations remains in doubt. Judging by the results from the aquaculture research and development projects conducted at the National Energy Laboratory of Hawaii, only those projects that have based their production strategy primarily on nutrient rather than temperature utilization of the deep ocean water have achieved some degree of commercial success. It appears that there is an important principle to consider before embarking on intensive aquaculture projects using OTEC water: that is, organisms that are primarily dependent on nutrients and photosynthesis,[1] such as micro- and macro-algae, or those that depend on the primary producers for food and growth, such as herbivorous fin fish and shellfish, are likely to be more successful commercially than those raised on artificial diets.

The ideal species for an OTEC-based aquaculture operation are likely to be those with relatively simple and well-studied life cycles. Species with complicated life cycles and exotic food requirements such as salmon, trout and abalone are difficult and costly to raise and may not be economically viable. The economics of OTEC-based aquaculture are also site-dependent. The commercial viability of an aquaculture operation relies in good measure on local market needs, social customs, consumer acceptance and economic returns. In essence, there is too little experience in OTEC-based aquaculture to predict with confidence its future success.

Other By-Products

In addition to generating power and aquaculture, OTEC and associated technologies can in some geographical locations provide fresh water, air conditioning and drip irrigation for specialized agriculture. If there is a local need for these by-products, the economics of an OTEC power plant operation can be significantly improved.

Although the upcoming deep cold water becomes warmer as it passes through the power generating section of the OTEC plant, it is still too cold for most aquaculture applications. Approximately one gallon of fresh water can be produced per degree of temperature rise for each 1,000 gallons of cold sea water flowing through the heat exchanger. Therefore, a 5 degree rise in the temperature of the exhaust from the OTEC plant, which would leave the cold water at a temperature suitable for aquaculture activities, could produce 5 gallons of fresh water for every 1,000 gallons of sea water passing through the heat exchangers.

The deep cold water that is pumped up could provide effective and relatively inexpensive air conditioning after it has gone through the OTEC power plant. If the plant is sited near a town, the cold water could be used for 'district air conditioning' in a similar manner as 'district heating' in Sweden and Iceland. The cold water, put through a heat exchanger, could offset approximately 20 times the electrical energy that an OTEC plant actually produces. Theoretically, a 1 MW plant could, in a future location, handle up to 20 MW of air conditioning load.

The use of fresh water condensing on pipes carrying cold sea water to grow non-indigenous plants is at an experimental stage. Experiments conducted at the National Energy Laboratory of Hawaii have pointed to the possible feasibility of growing strawberries and other temperate plants by managing temperature-induced seasonality.

Avery and Wu (1994) describe the potential of OTEC plants to produce methanol, synthesize ammonia and liquify hydrogen. These concepts could be implemented in a more distant future. Figure 12.5 shows the products that ultimately could be the result of OTEC commercialization.

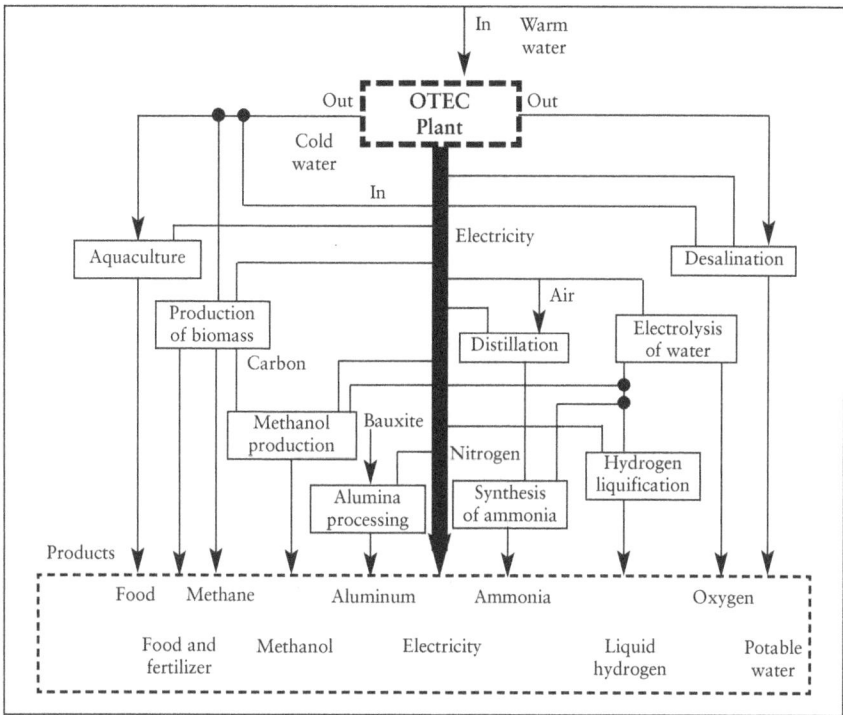

Figure 12.5 Potential Products of OTEC Commercialization
Source: Avery and Wu, 1994

The Environmental Impact

Commercial OTEC plants will have some negative effects on the marine environment. Construction facilities may disrupt the seabed, temporarily destroying marine ecosystems. Maintenance routines to reduce biofouling may also increase the level of toxic substances. If ammonia, freon, or some other environmentally inimical working fluid were spilled accidentally from a closed-cycle OTEC plant, the environmental effect on marine life in the immediate surroundings could be serious.

On the positive side, OTEC facilities release no additional heat and much less carbon dioxide than comparably-sized conventional fossil-fuelled power plants. Open-cycle OTEC produces more carbon dioxide than closed-cycle OTEC, yet the immediate carbon dioxide release from an open-cycle OTEC plant would be approximately 15 to 20 times smaller than the emission from a similar size fossil fuel power plant. Further, the release of carbon dioxide could be ameliorated if the discharge water is used for mariculture or other secondary operations.

For more than two decades OTEC has postured as a promising solution to the future supply of economic, reliable and environmentally friendly energy; but it has not yet fulfilled its promise. Using technologies derived from the offshore oil and gas industry OTEC requires further public and private sector commitment to succeed. The technological and financial capability to create OTEC plants rests in the industrialized countries in the northern belt, whereas the primary user market is mainly in tropical developing countries. Perhaps this is one of the underlying reasons why the growth of OTEC technology has not been hurried (Wright, 2002).

Prospects

From a technological viewpoint, the only problem OTEC presents is the large-diameter cold water pipe mounted on the sea floor for large (more than 2 MW) OTEC plants, or for plants mounted on floating platforms. In the long term, new pipe designs (soft and flexible) will be required. A major risk therefore is in the pipeline construction. A priority is to ensure good design and maintenance procedures, including the provision of an additional standby pipeline and pumping system which may mitigate the risk. Another important priority is to make OTEC more competitive by work on 'cost engineering' which would emphasize reducing the cost of pipe installation and construction (Tanner, 1995).

The success of OTEC rests on the reliable and economic operation of untested technology because an OTEC plant has not yet operated on a profitable commercial basis. Nonetheless, the engineering fraternity is confident that the technology, including the heat exchangers, will work well. An approach that could be applicable in the Caribbean is the use of man-made solar ponds onshore to provide the warm water, thereby increasing significantly the required temperature difference. In countries such as Saint Lucia, Dominica and St Kitts/Nevis which have geothermal potential, the geothermal field can be used to provide this high temperature water.

Various economic analyses of OTEC have concluded that the technology is feasible under selective conditions (for example, small tropical islands lacking fresh water) but it is at present approximately 40 per cent more expensive than conventional sources of energy under existing designs. This balance between OTEC energy and fossil fuels is changing rapidly as oil prices increase. As the cost of oil increases it should be borne in mind that OTEC would then be competing with other sources of renewable energy.

Early OTEC plants are likely to be built as joint ventures between governments and private industry on a negotiated arrangement. In this scenario, the private company builds the OTEC plant, bears the financial risk and sells power to the government-owned utility at a negotiated price.

From a strategic viewpoint it can be argued that Caribbean countries should participate in early OTEC plants, and adopt the approach of an entrepreneur in the international OTEC market. OTEC is a technology applicable to the tropics and there is a certain strong logic in a tropical country taking a leading role in the development of a tropical-related technology. The question is, can OTEC provide a profitable source of energy for the Caribbean island markets in the years to come? Indeed it can, but not before small commercial demonstration plants are built. The results from small experimental units (up to 1 MW) confirm the growing practicality of OTEC and the next requirement is the design, construction and operation of a representative scale denominator of say 5–10 MW, which will evaluate the feasibility of commercial OTEC production systems (Lennard, 1995). The second question is, can Caribbean countries afford to invest in the research and development phase of OTEC technology? The answer is yes, if one believes that OTEC is a technology whose time has come. Although OTEC cannot compete economically with present day fossil-fuelled electricity generation, it is clearly a potential energy source for Caribbean island states. Other ocean energy options, such as wave power, tidal energy and ocean osmotic energy should also be carefully studied by policy makers and energy-specialists in these countries.

Wave Power

Among different types of ocean energy, wave energy represents the highest density resource. Ocean waves encompass two forms of energy; the kinetic energy of the water particles, which in general follow circular paths, and the potential energy of elevated water particles. On the average, the kinetic energy in a linear wave equals its potential energy. The energy flux in a wave is proportional to the square of the amplitude and to the period of motion. The average power in long period, large amplitude waves commonly exceeds 40–50 kW per meter width of oncoming wave (Soerensen et al., 2008).

Several wave energy techniques are described below:

1. Oscillating water columns are partially submerged, hollow structures open to the seabed below the water line. The heave motion of the sea surface alternatively pressurizes and depressurizes the air inside the structure generating a reciprocating flow through a turbine installed beneath the roof of the device.

2. Overtopping devices, floating or fixed to the shore, that collect the water of incident waves in an elevated reservoir to drive one or more low head turbines.

3. Heaving devices (floating or submerged) mechanical and/or hydraulic convert up and down motion of the waves into linear or rotational motion to drive electrical generators.

4. Pitching devices consist of a number of floating bodies hinged together across their beams. The relative motions between the floating bodies are used to pump high-pressure oil through hydraulic motors, which drive electrical generators.

5. Surging devices exploit the horizontal particle velocity of the waves to drive a deflector or to generate pumping effect of a flexible bag facing the wave front.

The Agucadoura project in Portugal is the first large-scale commercial use of wave power. The wave power farm harnesses energy produced by ocean waves which is brought ashore by a submarine cable and fed directly into the national electricity grid. The first phase of the project will initially generate 2.25 MW of electricity using three Pelamis Wave

Energy Converters (PWEC); these are semi-submerged, articulated structures composed of cylindrical sections linked by hinged joints (Renewable Energy World, 2008). The second phase of the project will be to manufacture and install an additional twenty-five machines and take the installed capacity up to 21 MW. The generators are located approximately 3 miles from the coast. Initially the wave farm will provide electricity to 1,500 homes and when the second phase is installed, it can provide electricity to 350,000 homes. Notably, there are also pilot wave power projects in the Orkneys, west of Scotland, and Cornwall in southern England. These plants use a polanis system of flexible pipes, 3 metres in diameter and 100–150 metres long.

A new experimental device using a rubber tube named the "Anaconda" is being developed in the United Kingdom. It is simple in design with minimum maintenance and if it were commercialized, it could provide electricity at approximately 10 US cents per kWh, which is cheaper than other present wave power designs. In commercial operations, the Anaconda could be 200 metres long and about 7 metres in diameter, and deployed in water depths of between 40 and 100 metres.

The Anaconda is closed at both ends and filled completely with water. It is designed to be anchored just below the sea's surface, with one end facing the oncoming waves. A wave hitting the end squeezes it and causes a 'bulge wave' to form inside the tube. A bulge wave is a wave of pressure produced when a fluid oscillates forward and backwards inside a tube. As the bulge wave runs through the tube, the initial sea wave that caused

A Pelamis wave energy converter at the port of Peniche Portugal

it, runs along the outside of the tube at the same speed, squeezing the tube more and more and causing the bulge wave to get bigger and bigger. The bulge wave then turns a turbine fitted at the far end of the device and the power produced is transmitted to shore by a cable.

Because it is made of rubber, the Anaconda is much lighter than other wave energy devices (which are primarily made of metal) and dispenses with the need for hydraulic rams, hinges and articulated joints. This reduces capital and maintenance costs and scope for breakdowns. However, the Anaconda concept has only been proven at the laboratory scale and has to be scaled up to commerciality, a process that might take many years.

Tidal Energy

Tidal forces tend to be greater in regions of higher latitude. The tidal effect and hence the energy potential from tides is minimal in low latitude countries such as Jamaica. Tidal energy conversion techniques exploit the natural rise and fall of the level of the oceans caused principally by the interaction of the gravitational fields in the planetary system of the Earth, the Sun and the Moon. Tidal movement is diurnal in periodicity every 24 hours and semidiurnal every 12 hours 25 minutes. This motion is being influenced by the positions of the earth, moon and sun, with respect to each other. Spring tides occur when the tide generating forces of the Sun and the Moon are acting in the same directions. In this situation, the lunar tide is superimposed to the solar tide. Some coastlines, particularly deltas and estuaries, accentuate this effect creating tidal ranges of up to 15 metres. Neap tides occur when the tide-generating forces of the sun and the moon are acting at right angles to each other.

The vertical water movements associated with the rise and horizontal water motions termed tidal currents accompany the fall of the tides. Hence, there is a distinction between tidal range energy, utilizing the potential energy from the difference in height between high and low tides, and tidal current energy, the kinetic energy of the water particles in a tide or in a marine current.

Tidal currents have the same periodicities as the vertical oscillations, thereby being predictable, but tend to follow an elliptical path and do not normally involve a simple to-and-fro motion. Where tidal currents are channelled through constraining topography, such as straits between islands, very highwater particle velocities can occur. These relatively rapid

Tidal Energy Converters can be considered underwater wind turbines. Tidal waves slam into these propellers and much like wind turbines this rotational force creates energy

tidal currents typically have peak velocities during spring tides in the region of 2 to 3 metres or more (Soerensen et al., 2008).

Currents are also generated by winds, and temperature and salinity differences. The term "marine currents", often stated in the literature, comprises several types of ocean currents. Wind driven currents affect the water at the top of the oceans, down to about 600 metres. Currents caused by thermal and salinity gradients are normally slow, deep water currents that begin in the icy waters around the north polar ice. Wind driven currents appear to be less suitable for power generation than marine currents, as they are usually slower. Generally, tidal currents exhibit their maximum speed at fairly shallow waters, making them accessible for large engineering works.

The ultimate global tidal range energy potential is estimated to be about 3 TW, about 1 TW being available at comparably shallow waters. Recent studies indicate that marine currents have the potential to supply a significant portion of future electricity needs. The resource potential of the European marine current is estimated to exceed 12,000 MW of installed capacity (Soerensen et al., 2008).

Ocean Osmotic Energy

Osmotic power is a relatively new energy conversion concept even though osmosis has been known for several hundred years. All the basic technology components necessary for efficient osmotic power production are available in principle. New and more energy efficient membrane technology has been developed during the last few years.

The principle utilizes the entropy of mixing water with different salt gradients. In this process water is transported spontaneously through a semi-permeable membrane (i.e., a membrane that retains the salt ions but allows water through) from the freshwater side to the water with the higher salt concentration, creating pressure due to osmotic forces.

The increased pressure can be utilized in various forms, in this case to drive a turbine. Given a fixed volume compartment on the saltier side, the pressure will increase towards a theoretical maximum of 26 bars based on Atlantic sea water. This pressure is equivalent to a 270 metre high water column. It has been found that pressure retarded osmosis (PRO) is the most promising method for production of this energy. The principle of a PRO osmotic power plant is sketched in Figure 12.6.

Given sufficient control of the pressure on the salty water side, approximately half the theoretical energy can be transformed to electrical power, making osmotic power a significant new source of renewable energy.

In the PRO process, water with no or low salt gradient is fed into the plant (greyish) and filtered before entering the membrane modules. Such modules could contain spiral wound or hollow fibre membranes. In the

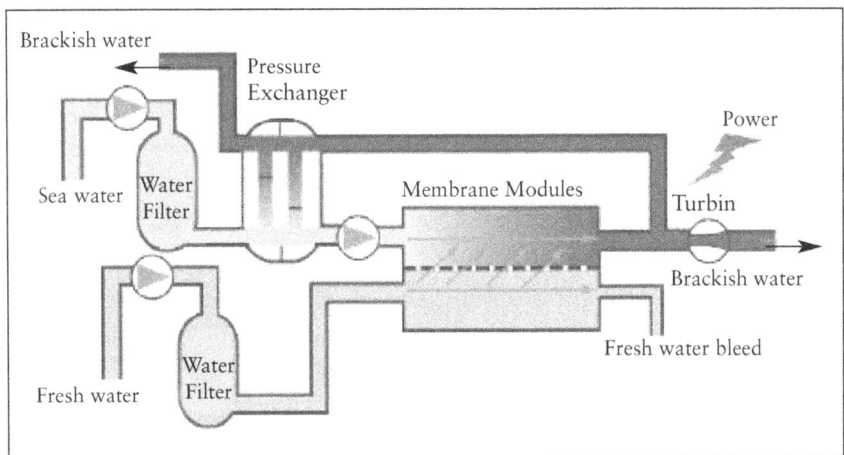

Figure 12.6 Principle of a PRO Osmotic Power

module, 80–90 per cent of the water with low salt gradient is transferred by osmosis across the membrane into the pressurized salty water (bluish). The osmotic process increases the volumetric flow of high pressure water and is the key energy transfer in the power production process. This requires membranes with particularly high water flux and excellent salt retention properties.

Economic Aspects and the Future of Ocean Energy

Apart from the fact that research from ocean energy is still nascent, there are at least five hurdles to be overcome before this technology becomes economical.

Determination of Costs. Due to the fact that the information on OTEC is relatively dated, it is difficult to provide an approximate estimate of the cost of osmotic-produced electricity. Another challenge is determining costs because of the large variety of cost estimates for reverse osmosis.[2] It is speculated that compared to other energy-producing processes, osmotic energy is about 36 times more expensive than a conventional power plant.

Electric Grid Access. Whereas ocean energy is coastally located, most electricity grids are central to a country. Usually transmission lines situated in the coastal parts of most countries are unable to accept the high loads of electricity that would be required to make ocean energy viable. As a consequence there will have to be significant upgrades in the quality of transmission facilities in these coastal areas. A major question therefore is how the grid expansion would be financed.

Regulatory Framework. Permitting any activity in a coastal environment creates difficulties, partly because of the lack of adequate baseline data relating to environmental factors including flora and fauna and the movement of currents. For this reason obtaining operational permits can be a long and difficult process. In many developing countries, too many agencies are involved in the permitting process without adequate coordination with each other.

Availability of Resource Data. Ocean energy development will find competition from alternative uses for the maritime space and marine resource. Thus the question of alternative uses for the maritime space and resource can result in delayed action and decision making by national authorities.

Economic Incentives. As a high cost technology, fiscal incentives and tariff arrangements will have to be developed that will spur investment and support high capital construction cost as well as consumer satisfaction.

The future of ocean energy lies in the long term. It is encouraging that a number of nascent small companies have an interest in ocean energy. Other advantages are the several attractive features in using salt for power, one of its most significant features being its renewability. There is no risk in running out of salt because of osmotic produced power.[3] The process of creating energy does not consume the salt but only utilizes it to force water to move. Another advantage is that osmotic-produced energy has minimal environmental impact, being a relatively clean process. The heat that is derived from the process would raise the temperature less than half a degree Celsius, which is not harmful to marine organisms. These advantages bode well for the growth in ocean energy over the ensuing two decades. The energy and environmental issues in which all countries are now seeking a panacea in wind, solar energy, nuclear energy, natural gas and a slowly emerging fuel cell technology will now be joined by the rising growth in new ocean energy systems. These emerging ocean energy technologies will hopefully mature and make a substantial contribution to the future global energy mix.

References

Avery, W.H., Wu, C. (1994) *Renewable Energy from the Ocean*. New York: Oxford University Press.

Lennard, D.L. (1995) "The Viability and Best Location for Ocean Thermal Energy Conversion Systems around the World". *Renewable Energy* 6 (3) pp. 359–365.

Renewable Energy World (2008) "Wave Energy to Power Portugal". Retrieved from: www.renewableenergyworld.com.

Soerensen, H.C., and Weinstein, A. (2008) "Ocean Energy: Position paper for IPCC". IPCC Scoping Meeting on Renewable Energy Sources ii – Proceedings.

Tanner, D. (1995) "Ocean Thermal Energy Conversion: Current Overview and Future Outlook". *Renewable Energy* 6 (3) pp. 367–373.

Wright R.M. (2002) "Will Ocean Thermal Energy Conversion Fulfill its Promise?". *Renewable Energy*. UK: Sovereign Press.

13. | Geothermal Energy
Utilizing the Earth's Heat

GEOTHERMAL ENERGY IS A VIABLE power alternative in many locations worldwide. Countries with suitable geothermal resources can use them to constitute a large proportion of the total energy resource base. The technology for using geothermal steam to generate electricity has existed since the early part of the twentieth century and has advanced steadily so that lower and lower temperatures of the resource have become adequate to power turbine generators (Dengo et al., 1993). An electrical geothermal plant can be sized to suit various situations. Approximately five megawatts is the smallest power load and additional generating capacity can be added in similar increments. This flexibility in size is not a viable option on a commercial basis for the standard thermal power plants.

Electricity is produced by geothermal in 24 countries.[1] Five of those countries obtain 15–22 per cent of their national electricity production from geothermal energy. The direct application of geothermal energy (for heating, bathing, etc) has been reported by 72 countries.[2] Presently, the worldwide use of geothermal energy exceeds 57 TWh/yr of electricity and 76 TWh/yr for direct use (Fridleifsson, et al., 2008). It is considered possible to increase the installed world geothermal electricity capacity from the current 10 GW to 70 GW with present technology and to 140 GW with enhanced technology. Enhanced Geothermal Systems, which are still at the experimental level, have enormous potential for primary energy recovery using new heat exploitation technology to extract and utilize the Earth's stored thermal energy.

Geothermal Resources and Technologies

Geothermal resources can be partitioned into four categories:

- Hydrothermal
- Geopresssured geothermal
- Hot dry rock[3]
- Magma

Currently, all commercial operations are based on hydrothermal systems where wells are usually between 1,800 and 2,450 metres deep with reservoir temperatures of 180°C –270°C. The Geysers electric power geothermal field in California (1,850 MW) is perhaps the best known. Geopressured, hot dry rock systems and magma are still experimental, although the former two have been technically demonstrated successfully and energy generation verified. Hot dry rock technology development has been conducted in the USA, Japan, France, Germany and the United Kingdom. Technology using hot rock could extend development of geothermal resources into many countries. The growth in the use of hot dry

rock resources will depend on gaining access to water supplies for injection into the hot dry rock. The geographic distribution of possible magma resources is not well known at this time, but prospects are best where there is volcanic activity. The technology does not exist to handle the working environment of magma which has temperatures of between 650°C and 1,300°C.

Three primary conversion technologies can be used for power generation from hot water (hydrothermal):

1. **Flash Steam Plants** (for resource temperatures more than 175°C) which rely on flashing the hot water to steam. As shown in Figure 13.1, single-flash systems vaporize hot geothermal fluids into steam which drives a turbine. In a dual-flash system, steam is flashed from the remaining hot fluid of the first stage and passed into a dual-inlet turbine or into two separate turbines. Condensate can be used for cooling and the brine returned to the reservoir. This technology is used in many locations including Japan, New Zealand, Mexico and the Philippines.

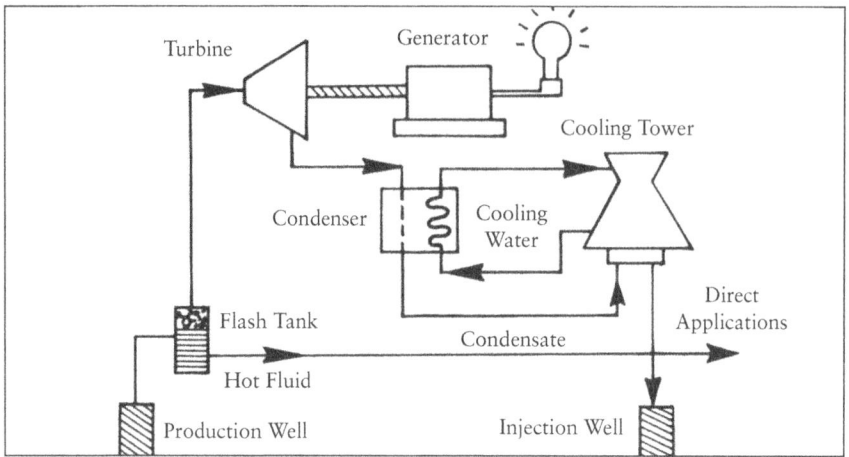

Figure 13.1 Flash Steam System

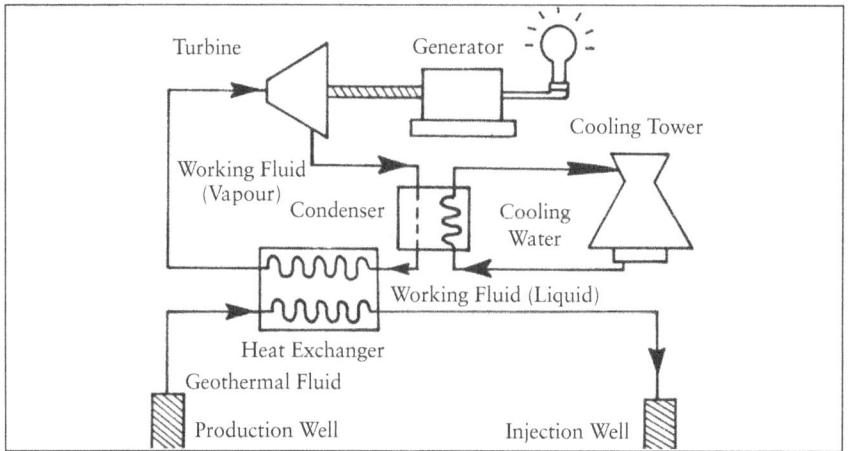

Figure 13.2 Binary Geothermal Systems

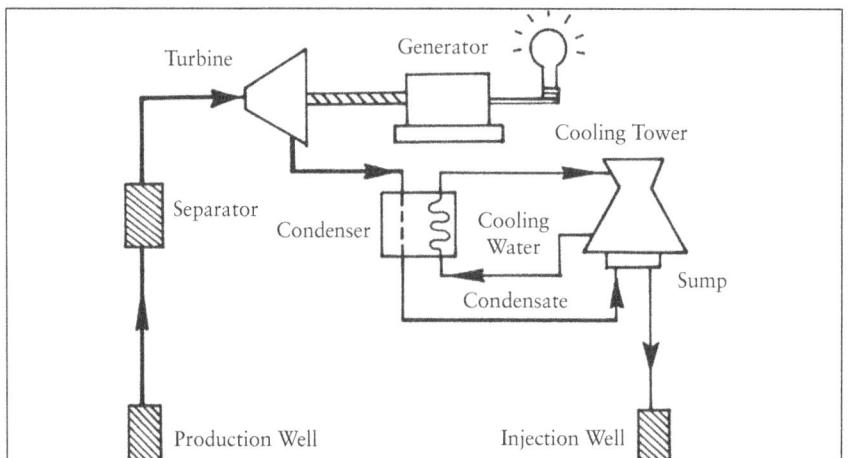

Figure 13.3 Dry Steam Geothermal System

2. **Binary-Cycle Plants** (for resource temperatures 100°C–175°C) use the hot water to boil a working fluid (an organic compound) which drives a turbine. This system can also make use of the heat remaining in water separated from steam in flash plants (for example, Kawerau, New Zealand). These plants are also useful for saline brine that cannot be flashed because of the resulting deposition of scale, and also in the case of geothermal fluids with a high dissolved (non-condensible) gas content.

   In this process, the hot geothermal fluid is maintained under pressure by a down-well pump and passed through a heat exchanger where it vaporizes a working fluid such as isobutene. The working fluid, expanded as vapour, is moved through a turbine, condensed, and reheated for another cycle (Figure 13.2).

   Binary systems tend to have higher conversion efficiencies than flash steam plants and need lesser quantities of geothermal fluids per unit of electricity generated. Because there is no steam condensate in the process, binary plants demand external sources of cooling either by water or air. Binary cycle technology is a prime choice for use with hot dry rock resources because of the moderate temperatures of the fluids circulated.

3. **Dry Steam Plants** (Figure 13.3) are used to produce energy where the reservoir is mainly vapour. Steam is gathered from the well, cleaned to remove entrapped solids, and moved directly to a turbine (California Energy Commission, 1988).

   From an engineering perspective, the geothermal fluid is corrosive, and is at a relatively low temperature (compared to steam from boilers), which by the laws of thermodynamics, limits the efficiency of heat engines in extracting useful energy as in the generation of electricity. Much of the heat energy is lost, unless there is also a local use for low-temperature heat, such as greenhouses or timber mills or district heating.

Economic Issues

From an economic view, geothermal energy is extremely price competitive in some areas and reduces reliance on fossil fuels and their inherent price unpredictability. It also offers a degree of scalability: a large geothermal plant can power entire cities while smaller power plants can

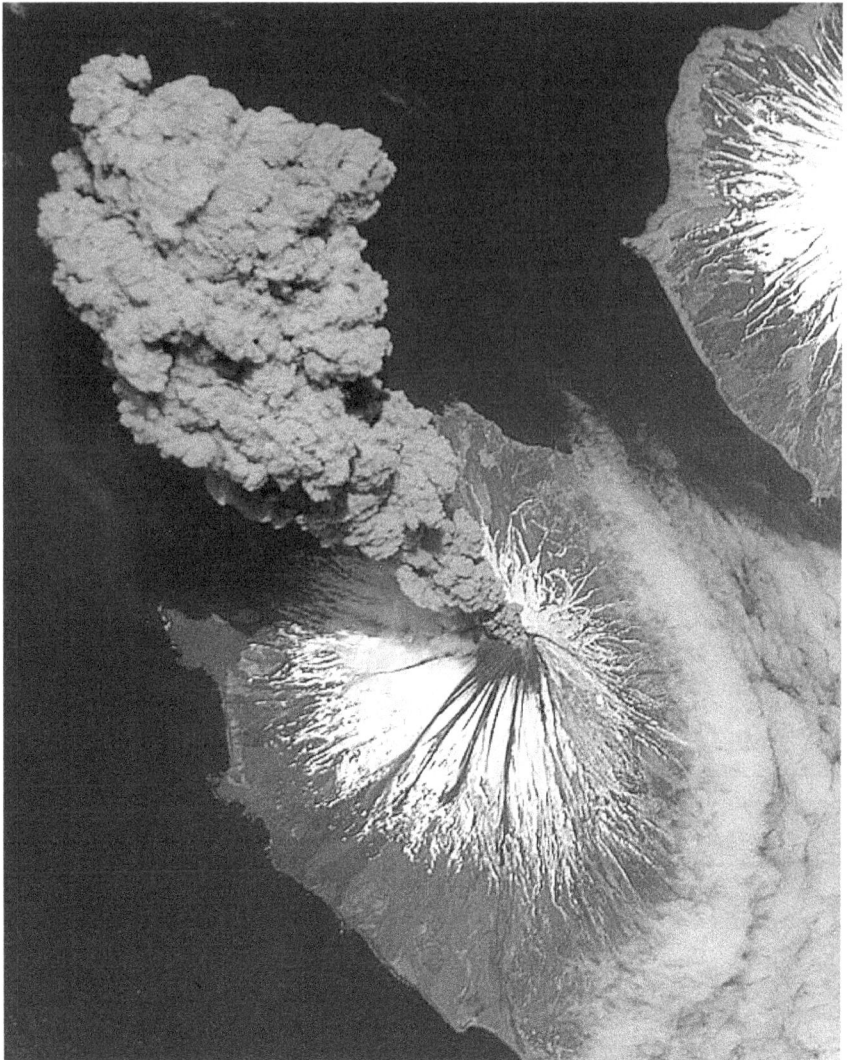

Alaska's abundant geothermal sources provide hot water for domestic use

supply more remote sites such as rural villages. Moreover, geothermal plants are more reliable than those fired by fossil fuels. The plants usually need to be taken off-line for maintenance and repair only about 5 per cent of the time, while coal plants, for example, are off-line for about 20 per cent of the time. Geothermal energy offers another advantage over traditional fossil fuel based sources, primarily because the heat source requires no purchase of fuel. In addition, geothermal power plants are unaffected by changing weather conditions and they also work continuously, day and night, making them base load power plants.

The current investment cost in geothermal energy is approximately US$3.5–5.5 million/MW installed. The direct use of geothermal energy for heating is also commercially competitive with conventional energy sources. Geothermal energy is available for 24 hours every day of the year, and can therefore serve as a supplement to energy sources which are only available intermittently (Fridleifsson, op. cit.).

Environmental Issues

If properly managed, geothermal powered plants are relatively benign where emissions are concerned; the plants release much less carbon dioxide than generating plants powered by fossil fuels. Geothermal energy is nearly sustainable because the heat extraction is small compared to the size of the heat reservoir, which may also receive some heat replenishment from greater depths. Despite the advantages, there are several environmental concerns related to geothermal energy. The construction of the power plants can adversely affect land stability in the surrounding region. This is mainly a concern with Enhanced Geothermal Systems, where water is injected into hot dry rock where no water was before. Dry steam and flash steam power plants also emit low levels of carbon dioxide, nitric oxide, and sulfur, although at roughly 5 per cent of the levels emitted by fossil fuel power plants. Geothermal plants can be built with emissions-controlling systems that can inject these substances back into the earth, thereby reducing carbon emissions to less than 0.1 per cent of those from fossil fuel power plants. Hot water from geothermal sources will contain trace amounts of dangerous elements such as mercury, arsenic, antimony, which if disposed of into rivers can render their water unsafe to drink.

Although geothermal sites are capable of providing heat for many decades, specific locations may eventually cool down. It is likely that in these locations, the system was designed too large for the site, since there is only so much energy that can be stored and replenished in a given volume of earth. Some researchers interpret this as meaning a specific geothermal location can undergo depletion, and question whether geothermal energy is truly renewable.

The Jamaican Situation

Electrical-grade geothermal energy is concentrated within active volcanic regions. Volcanoes are indicators of near-surface crustal magma that forms the source of heat for hydrothermal-convection systems at drillable

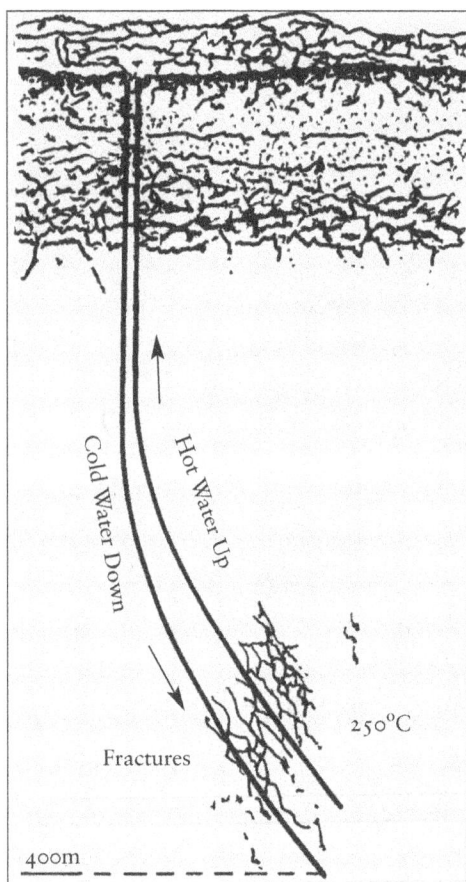

Figure 13.4 Hot Dry Rock Geothermal Concept

(shallow) depths. Therefore, geothermal energy is related to volcanism, plate tectonics and fault movements. Extensive volcanism occurred in Jamaica in the lower and upper Cretaceous (about 100–70 million years ago) with the youngest phase of volcanism taking place in the Pliocene period (8 million years ago) on the northeastern coast of the island. There are no existing major heat sources at shallow depths within the crust of the island. The Plaintain Garden Fault in the Blue Mountain Block, and other related fault systems therein, make the Blue Mountains the only area of possible geothermal potential. Hot water springs occur in many localities in the Blue Mountain Block, not only along major faults (for example, Bath Springs) but also in association with minor faults within the Block.

Further investigation into geothermal energy in Jamaica would probably include drilling a well in the vicinity of Guava River Springs in the Blue Mountain area of Portland to better understand the geothermal gradient there. In general, Jamaica's geothermal energy resources are in the form of low-enthalpy (low temperature) waters suitable for heating (no demand) and not of a grade suitable for generating electricity. The geothermal possibilities in Jamaica lie in hot dry rock technology to form a man-made geothermal reservoir in low permeability rock with hydraulic or explosive fracturing. In this case, injection and projection wells could be joined to form a circulating loop through the man-made reservoir, and water then circulated through the fracture system (Figure 13.4).

New Developments in the Eastern Caribbean

With oil prices spiralling out of control at least one Caribbean island is pushing to begin producing geothermal energy to meet a hundred per cent of its electricity needs by 2012, as well as for exportation to neighbouring Caribbean islands. With its population of 12,500 covering some

36 square miles, Nevis will begin production of the renewable energy by 2012. The island has the potential to produce as much as 900 megawatts of electricity from geothermal sources.

Dominica also intends to enter the Geothermal energy industry. A licence has been granted by the Government of Dominica to West Indies Power Dominica Limited (WIPD) which will develop Dominica's geothermal energy resources in the south of the island. This is in addition to the preliminary exploration of the geothermal energy resources in the Wotton Waven area in Dominica under the Geothermal Energy in the Caribbean Islands project, funded under the Interreg III B Programme of the European Union (Government of Dominica, 2008). Other Eastern Caribbean islands with important potential geothermal reserves include Saint Lucia and Montserrat.

Future Trends

Geothermal energy is limited to a narrow niche market in electricity generation. Where hydrothermal resources exist, geothermal energy can compete with thermal or internal combustion power plants. Trends for geothermal energy are the following:

1. The price of petroleum will affect the business climate for geothermal resource development and the pace at which private capital is invested as well as government support for research and development.

2. When environmental costs are internalized, renewables such as geothermal will have economic advantages. Nonetheless, geothermal plants are not always considered to be environmentally benign. The uncertainty of the hydrothermal reservoir performance is still a problem, and in general, technological improvements are needed in the geothermal industry.

3. The growth of the geothermal market is not hindered by a paucity of resources; 11,000 MW of proven resources are yet to be developed. Indonesia and the Philippines are areas slated for growth in geothermal energy production. Hot dry rock and geopressured technologies should become mature by about 2015.

Scenarios for future development show only a moderate increase in traditional direct use of applications of geothermal resources but a rapid increase is expected in the heat pump sector because geothermal pumps can be used for heating and or cooling in most parts of the world. There is great potential for the enlarged use of geothermal as a part of the global energy matrix. The technology is independent of weather conditions and given its inherent storage capacity it is particularly useful in providing base load power. Geothermal has limited environmental impact as compared to some other energy resources. Combined heat and power plants have seen growth, and the overall efficiency of geothermal energy utilization has been improved.

References

California Energy Commission (1988) Energy Technology Status Report. Report P500-88-003.

Dengo, G., Duffield, W.A., and Heiken, G.H. (1993) "Geothermal Steps towards Independence in Central America". *ACID News*, No. 7475, pp. 19–22.

Government of the Commonwealth of Dominica (2008) "Nature Island of Dominica Now in the Vanguard of International Efforts for the Development of World's Renewable Energy Resources". Retrieved from: http://www.dominica.gov.dm/cms/index.php?q=node/419

Fridleifsson, I.B., Bertani, R., Huenges, E., Lund, J.W., Ragnarsson, A., and Rybach, L. (2008) in Hohmeyer, O., Trittin, T. (Eds). The Possible Role and Contribution of Geothermal Energy to the Mitigation of Climate Change. Luebeck, Germany. pp. 59–80. Retrieved from: http://iga.igg.cnr.it/documenti/IGA/Fridleifsson_et_al_IPCC_Geothermal_paper_2008.pdf.

14. | Sustainable Architectural Design
Reducing Energy Use in Buildings

SINCE THE EARLY 1990S, EFFORTS to improve the energy performance of residential, commercial and institutional buildings have been increasing. The sun is an important factor in keeping a building warm or cool, providing both assets and liabilities. The promises of economic, direct solar energy conversion to electricity by photovoltaics and other solar technologies encourage the concept that urban buildings will be able to produce much of their own energy needs. This transformation of the city from 'consumer' to 'producer' of energy is a potential enterprise of enormous magnitude (Hawkes, 1995).

If buildings could be clad in advanced materials and able to utilize technological systems capable of converting more natural energy, we could have an ultimate paradigm – the sustainable city. To achieve a sustainable city, we need to change our architectural designs from air-conditioned glass-walled buildings that consume energy. With the ultimate goal being the sustainable city, a number of measures can be addressed immediately.

Sustainable building and design can be described as environmentally friendly, environmentally conscious, energy conscious, climate conscious, bioclimatic, energy-efficient, low energy, greener or simply green architecture. There is really no internationally agreed definition for green or sustainable development. Sustainable development involves energy efficiency as well as efficiency in the use of water and materials, the provision of high indoor air quality and the reduction of wastes. This entails harnessing renewable natural resources such as solar energy, and utilizing

materials that cause the least possible damage to the global commons – particularly water, soil, air and forests.

Sustainable building encapsulates various combinations of response to climatic conditions. Sustainable building also improvises blends of traditional and innovative building techniques, appropriate material selection, implementation of low energy building systems, and where possible the use of renewable energy systems as viable alternatives to powering buildings.

A multi-pronged approach is required to stimulate the building industry to design and build to levels of performance that are appropriate for a sustainable agenda. Key elements include regulations, demonstrations, financial incentives, support tools and performance labelling programmes for equipment such as air conditioners. The major energy 'flows' through a building are of energy, water and materials. The basic strategy is to:

- increase energy efficiency
- use less water
- use materials efficiently
- use recycled and renewable sources of materials
- select materials with the lowest environmental impact

In Jamaica, unrestrained urbanization with unplanned building arrangements has negatively affected valuable natural resources of energy, water and ground cover, hampering the process of eco-friendly habitat development. Urbanization in Jamaica is characterized by expanding patterns of growth, often concentric, consuming land in the immediate proximity of cities. Increased efficiency in energy use offers the most significant and cost effective approach to reducing environmental degradation. There is also the need to accelerate the development and market penetration of efficient and sustainable energy technologies.

Sustainability is important for the building sector in Jamaica because:

- approximately 50 per cent of material resources taken from nature are building related
- more than 50 per cent of national waste production is sourced from the building sector
- about 30 per cent of the energy consumption is building-related, as shown by sector energy use studies conducted by the Petroleum Corporation of Jamaica

The main issues in sustainable architectural design are:

- comfort, influenced by the age and activity level of the occupants. Comfort includes aspects of the internal environment such as air, temperature, noise and light
- health, as it relates to indoor air quality, materials, noise and the transmission of communicable disease
- the environmental impact of the building itself. Buildings are users of energy, materials, water, and producers of waste and noise. Buildings contribute to larger impacts such as global warming and the depletion of natural resources

On the micro scale, the objective is to shelter occupants from the elements, maintain a comfortable thermal environment and provide visual well-being and adequate ventilation. These objectives can be achieved by many strategies, including but not limited to:

- locating buildings in relation to natural topography and prevailing winds
- making structures and roofs resistant to high wind speeds
- taking measures to prevent local flooding
- providing an ambient temperature, giving thermal comfort to occupants (23°C–26°C are reasonable). The optimal temperature should be achieved at knee height (0.5 m from the floor)
- having cooling systems that are easy to control
- designing an average ratio of window to floor area of 15 per cent for the building as a whole as a good starting point

- having shading devices such as screens, blinds, and photo thermo electro-chromatic glass to aid in controlling sunlight penetration
- using light-coloured paint on external walls to reflect solar radiation.
- designing orientation and correct spacing of buildings to enhance natural lighting
- making the operable area of windows to extend close to ceilings in order to allow hot air in the upper part of the room to escape
- having ventilation rates comply with air quality standards and sanitary recommendations; average of $25m^3$ per person per hour (in offices)

The microclimate is central to the relationship between urban forms and spaces and strategic energy-efficient design. The microclimate is dictated by temperature, wind patterns, sunlight and air quality. The macro scale considers land use, density and urban transportation. Affordable road transport and specialized land-use zoning have encouraged dispersed settlement patterns in most countries, requiring citizens to take long journeys. The private motor vehicle is a wasteful user of energy and a significant source of emissions, so sustainable design should be combined with measures to reduce its use. There should be effective public policies for traffic flow, parking restrictions, road pricing, bus/taxi priority.

Design and density are intertwined. Higher densities result in lower energy consumption in buildings, less wastage of Greenfield sites and more use of public transport. In fact, the commercial survival of many services depends on relatively high population densities for its customer base.

Energy efficiency is affected by density, attention to land-use, transportation, water and waste. Architects should ensure that each new or refurbished building minimizes the energy embodied in its construction. Architects should also use environmentally friendly and economic sources such as thermal and photovoltaic solar systems, wind, and small hydropower, together with cogeneration units. Cogeneration, an energy efficient technology, is pertinent to industry, commerce and larger hotels.

One of the basic tenets in using renewable energy is to utilize technologies and ways of working that do not harm the environment and favour sustainable development. Renewable energy projects make a decisive contribution to this objective in different ways. The project developers should make every effort to avoid the emission of greenhouse gases to the atmosphere, a major factor in climate change.

A 20 MW wind farm at Wigton in Jamaica[1] avoids the emission of approximately 50,000 tonnes of CO_2 per annum. The methodology used in the selection of the wind farm site is highly respectful of the environment. The wind farm is more than 200 metres from housing and is sited on lands to be used for bauxite mining, allowing a multipurpose use of the land. Efforts were made to ensure that the impact on the environment during construction was reduced to a minimum. Existing roads and tracks were used wherever possible, and excavated earth used to recover the vegetation on the land.

Buildings should be designed to meet the occupant's needs for thermal and visual comfort at reduced levels of energy and resources consumption. Energy resource efficiency in new construction can be realized by adopting an integrated approach to building design. The four main steps in this approach are stated below.

1. Use passive solar techniques in building design to reduce load on conventional systems (cooling, ventilation and lighting) – this involves the use of natural ventilation, reduction in solar gains, the use of sky, wet surfaces, and vegetation.

2. Design energy-efficient lighting, ventilation and air conditioning systems.

3. Use renewable energy systems (solar photovoltaic systems/solar water heating systems) to meet a part of building load.

4. Use low energy materials and methods of construction and reduce transportation energy: the use of material with low embodied energy forms part of energy efficiency building designs.

Solar Energy

Photovoltaics provide an electricity option for solar energy in Jamaica, both in remote areas where it is too expensive to build transmission lines, and in urban areas where it can be used on rooftops. The ultimate advent of net metering in Jamaica will spur growth in this technology, because it will allow homes to sell electricity to the grid in the daytime and buy from the grid at night, all through the same meter. The facade integration of photovoltaics will become increasingly familiar, while solar thermal collectors mainly on rooftops will provide more and more hot water production.

Facade collectors, whether photovoltaics or solar thermal are energy converters, and are an integral part of architecture. Rooftop mounted solar collectors are important, and architects and planners should cooperate in building projects in order to give the solar collectors aesthetic value. Indeed, a barrier to the use of rooftop mounted solar collectors in the Jamaican market is that many customers consider them to be a visual blight.

Solar Heating

Energy for buildings accounts for an important part of Jamaica's energy requirements. The best known domestic use of solar energy is for heating water. Solar heaters have no adverse environmental effect at any time in the life of the system.

According to Szokolay (2002), solar heat gain depends on

1. irradiance of the window plane
2. the glazing material
3. any external shading devices
4. any internal controls (blinds, curtains, etc)
5. interior surfaces and materials
6. indoor–outdoor temperature difference

The use of low-iron tempered glass, improved insulation and durable selective coating has led to increases in the performance of solar heater collectors from the 35 to 45 per cent of the early 1970s to the 50 to 60 per cent range available today. Major improvements have been made in the reliability of solar heaters with the advent of brushless pumps for solar systems although temperature sensors still seem susceptible to drift and failure.

Quality is an important requirement of a successful solar water heater. The market for solar heating will expand in developing nations when affordable systems of high quality are available and when users have reason for complete confidence and satisfaction in them. Solar heating systems have been granted tax incentives by the Government of Jamaica to stimulate their increased use. It is now simply a question of stimulating the market.

Solar Cooling

Active solar cooling may be defined as utilizing the sun's energy to help offset the net cooling load (space-conditioning or refrigeration) of a building. The concept of solar cooling makes sense because the cooling load is roughly in phase with the sun's intensity.

The coupling of solar energy and the absorption refrigeration concept to produce cooling was first attempted in 1872 in France when Abel Pifre used steam from a solar-heated boiler to operate an absorption cooler. Early units operated intermittently, extracting heat at night from a substance which was cooled and then replenished through solar heating during the following day. These devices were generally used to produce ice to preserve food.

By the 1960s, continuously operating solar-powered cooling systems suitable for either food preservation or air-conditioning had been demonstrated. Various experimental heat-driven Rankine, desiccant, and absorption systems were developed and field tested. The high initial cost and poor thermal performance, however, prevented solar cooling from being speedily implemented.

The economics of solar cooling, not unlike many other solar technologies, depend on several fluctuating elements, among which are:

- interest and inflation rates
- capital cost of solar and conventional equipment
- taxes and depreciation
- operating costs, including fuel rates

The performance of solar-actuated chillers and dehumidifiers has improved by approximately 50 per cent over the past ten years. It is still necessary to increase system efficiency as well as to reduce the size and cost of the ancillary collector array. Improvements in the performance of these systems will ultimately come about through better solar collector design, increased chiller component heat-transfer effectiveness, new liquid and solid desiccant materials, and more effective cooling system configurations. The reliability of cooling systems and collectors will be important issues. Not until cost, performance and reliability are resolved will solar cooling systems be commercially viable. At present, costs remain high and the industry is stagnant.

Passive Cooling

Passive cooling is a low energy-intensive method of keeping a building cool by relying on architectural design. Passive cooling reduces the auxiliary energy load by ventilating, cooling and providing lighting and so reduces the operational costs of the building. Heat avoidance techniques, natural lighting and natural cooling methods are incorporated in the structure to minimize energy consumption while improving the indoor comfort level. Natural cooling methods include natural ventilation, night cooling, earth-contact cooling and cooling by evaporation or radiation. Passive cooling systems often utilize the same building materials found in conventional structures, operate with little or no mechanical assistance and are very unlikely to malfunction. For many decades, buildings in the tropics have used passive cooling techniques to optimize comfort levels. Among these are buildings raised above the ground on blocks or stilts to improve ventilation, covered porches, large overhangs and light-coloured roofs to reflect the sun. Other key aspects of passive cooling technology include the use of insulating materials that retard heat flow, air infiltration, radiant heat transfer barriers below the roof, window design, desiccants for moisture reduction and new types of high performance glass.

The benefits of passive cooling are obvious: considerable peak load reduction for the utility company, improved comfort, lower utility bills and little additional cost to the builder. An important component of passive cooling is the siting of the building. The ideal bioclimatic layout of tropical buildings is low density in dispersed patterns with access to ventilation on all sides by porches and verandahs. This ideal is no longer possible due to pressure of population on land resources. The resulting high cost of land means that the distance between buildings is often reduced to a minimum. As Cooke (1980) shows, lower building densities have many advantages including:

- maximum flow of air around all sides of the building avoiding hot spots and solar-heated recesses

- allowing the use of outdoor space as an alternative or complement to indoor functions
- providing privacy by distance so that air movement is not impeded by walls

Placing tropical buildings on stilts above ground and extending their vertical massing are techniques of maximizing access to cooling breezes. Orientation is also important in site planning. Solar-gain surfaces should face the sun to optimize irradiation. All other surfaces should be shaded. Also, properly shaded structures should be, as far as possible, wind oriented.

There is a variety of passive ventilation techniques such as solar chimneys, Trombe walls, wind towers and roof vents. Domed roofs resist solar gain and improve ventilation (Gadi and Ward, 1990). Givoni (1994) further suggests the use of pumps as part of the passive concept which enhances a building's indoor climate.

The limiting factors in heating or cooling in a given climate are:

1. **Thermal Mass:** the material with which the buildings are made should preferably be indigenous. Materials of a region are usually the most suitable building material for that location and climate. In woodland, wood is the obvious choice for building material; in a mountainous region, stone might be preferable. A good building material should have a thermal dispersion that is appropriate to the region. If we require cool outside air to penetrate the building and reduce the temperature then the building should be designed with the shortest possible thermal dispersion, say half an hour. A high mass building (concrete walls externally insulated) will have significantly lower maximum indoor temperatures than one with a low mass. Therefore, for light-weight buildings in the tropics, continuous ventilation is the main comfort strategy (Givoni, 1994).

2. **Ventilation:** one of the simplest methods of achieving comfort in a building, especially if cross-ventilation is coupled with good indoor air speed. This type of cooling is the best strategy in warm, humid regions, with air speed between 0.5 and 2 m/s, when the outside air temperature does not exceed 32°C and the diurnal range is less than 10°C.

3. **Night Ventilative Cooling:** convective cooling of the building structure applies when the diurnal temperature exceeds 12°C and works best when the minimum summer temperature is below 20°C. In such cases

the coolness of night air can be stored within the structural mass of a building with air speed of 2–3 m/s or exhaust fans can be used to remove the hot air from the building and replace it with cool night air.

4. **Radiative Cooling:** radiative loss during the night can cool roofs, walls and floors by an average of 3°C to 5°C during the night and be used during the day to gain interior comfort. Thick roof construction is one approach and even water ponds on the roof can be used for cooling.

5. **Shading:** this combines little or no solar gain with retention of coolness from night radiation within the shaded zones. There are many designs of windows which reduce solar radiation transmittance into buildings (Taleb & Al-Wottar, 1988; Talmatamar et al., 1955).

The first step in shading is to orient the building to have the least direct impact from the sun. The next step is to prevent solar gain by means of shading devices, rejecting the solar gains by means of ventilation and absorbing solar gain in thermal mass. Wall colour can be important; a white wall, or a wall shaded by vegetation is effectively exposed to a low level of radiation, even when facing east or west. Eastern windows equipped with appropriate operable shutters can be protected from the sun while taking advantage of ventilation from the easterly wind (Givoni, 1994). Fixed external shading devices such as awnings are effective because the solar radiation is rejected before it enters the building. This effectiveness varies with the seasonal geometry of solar radiation, as affected by the sun's movement. Indoor shading devices reflect solar energy which has passed through the glazing into the room and back out through the glazing. They are less effective than external shades because some radiation is reflected back into the room and some absorbed by the surface of the shading device. All shading devices influence the view through glazed windows: an overhang, a Venetian blind, an opaque blind, fine wood lattice, bamboo screens and solar control films may all reduce the solar gain by the same amount but they will alter the view through the aperture quite differently in each case.

Evaporative Cooling is applicable mainly in arid regions, this process is adiabatic with no heat being gained or lost. The air is cooled by mechanical evaporative cooling. Evaporative cooling can be produced by the use of porous roof materials (Kimura, 1994). The roof materials soaked by rain water are naturally evaporated by solar radiation. The roof materials used traditionally include thatch, wood shingles, wooden skin and unglazed roof tiles. They are all moisture-absorbent, allowing evaporative cooling, while glazed roof tiles are not.

Day Lighting

Day lighting involves the skilful use of natural light as an effective lighting source for a building during its daytime operation. According to De Herde and Nihoul (1994), day lighting design approaches include four concepts:

- penetration: collection of natural light inside the building
- distribution: homogenous spreading of light into the spaces or alternatively focusing of the light
- protection: reducing the direct penetration of the sun's rays into the building by shading
- control: controlling light penetration by fixed or moveable screens to avoid visual discomfort

The strategies for day lighting, cooling and heating are illustrated in Figure 14.1.

An interest in day lighting emerged in the 1970s with the awareness that in many large buildings the single largest energy consumer was electric lighting and not cooling. However, energy savings do not result if electric lighting cannot be turned off or dimmed. Day lighting has merits beyond mere energy savings. It is extremely effective at reducing peak load which will lower peak cost to the utility. The building owner benefits from reduced consumption whilst the utility benefits by being able to spread out demand in order to lower peak cost. This could help to reduce the long-term need for new power plant construction. It may be well to consider providing direct financial incentives for day lighting strategies in order to reduce demand.

The most significant technological improvements have been in the availability of new glazing materials. One example is selective transmitting low-emissivity (low-e) glass coatings which block unwanted heat while allowing in visible light. Another advance is in the area of optical switching materials (i.e., thermochromatic, photochromatic and electrochromatic glass) which control their optical properties in response to light, heat or an electrical field.

The environmental function of a building is to mediate between the variable external climate and the stable conditions required for human

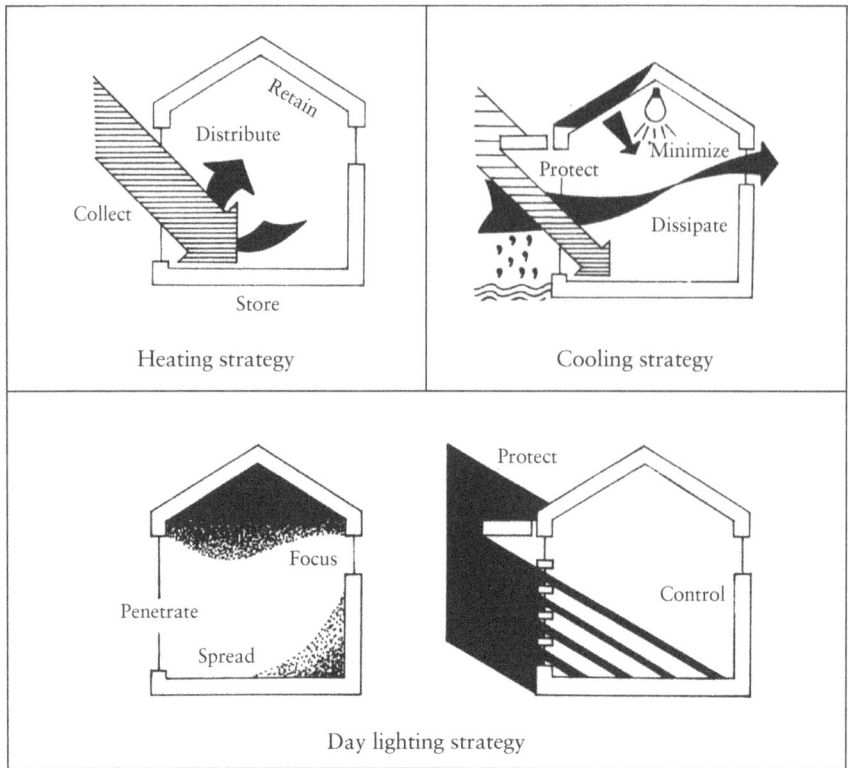

Figure 14.1 Strategies for Heating, Cooling and Day Lighting in Buildings
Source: De Herde and Nihoul, 1994

comfort. Bioclimatic awareness in urban planning, day lighting, and natural cooling of buildings results in lower energy consumption and a more benign effect on the physical environment.

The Atrium

In sustainable architecture, a number of approaches are pertinent. An important architectural style for both lighting and cooling is the concept of an atrium within a building. The atrium can be used to achieve four lighting objectives at increasing levels of difficulty. Perhaps the simplest of these is providing sufficient levels of illumination for occupants within the atrium itself. Usually, the designer will aim to control direct sunlight so as to prevent glare and overheating. From a psychological standpoint, occupants should have lighting arriving from as many directions as possible and if this is unattainable through natural sources, an element of balancing by use of artificial lighting may be necessary. Solutions to the objectives of design for lighting and cooling should be developed in tandem, as they have the potential to contradict each other. Another requirement is the provision of sufficient illumination to maintain plant growth within the atrium. This should be an integral part of the initial design process. Plants require adequate amounts of illumination, to provide them with energy, to create seasonal guidance and to regulate their growth and shape. Daylighting remains the element of plant maintenance that is most difficult to predict and control in the interior environment.

Examples of Sustainable Architecture

In low-energy architecture, the use and control of wind is an important factor in the design of a bioclimatic house. One example is the wind towers used in the desert climate (hot days and cold nights) of Iran. In the Middle Eastern countries, wind towers with evaporative cooling pads placed at the tower entrance are a highly effective passive design feature. The structure is a kind of chimney divided into vertical sections by brick walls. At night the tower cools off. During the daytime the walls cool the surrounding air and, as air becomes denser, it flows downward and into the building where the pressure of the cool air pushes the hot air out (Figure 14.2). Heat stored in the wall during the day is passed to the air at night, creating a rising current. This is an ancient system of air-conditioning with variations found in buildings in Italy and many

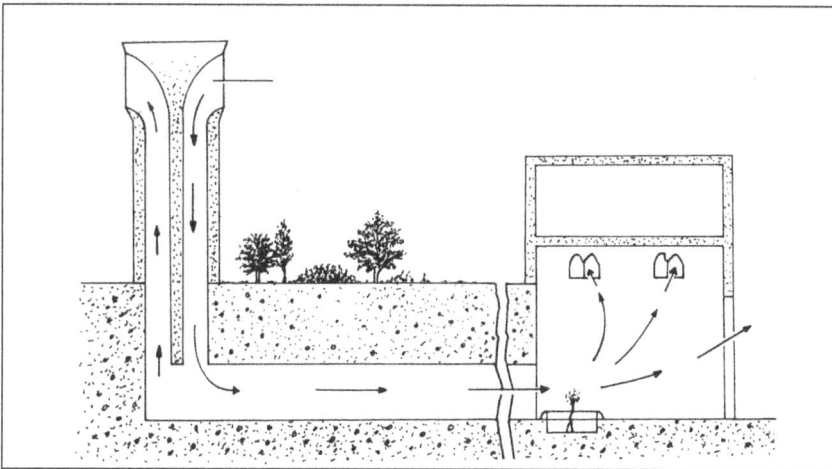

Figure 14.2 Principle of a Wind Tower and Cooling Chimney. Cool air enters the tower on the windward side and its pressure pushes the hot air out of the building
Source: Gallo, 1994

Mediterranean countries, dictated by the need to provide interior comfort before elaborate artificial systems came into use. Wind towers, with an element of evaporative cooling from moist surfaces, may well find practical application in the Caribbean. Conventional and modern versions of wind towers can be incorporated aesthetically into the design of modern buildings.

The primary purpose of passive solar designs is to reduce commercial energy consumption. Such designs need to be managed effectively after occupancy if potential benefits are to be realized (Nutt, 1994). Figure 14.3 sets out a matrix for the phases for important decision making. It can provide a structure for describing and comparing the energy impacts of pre-occupancy design and engineering decisions with the potential impacts of post-occupancy use and management. It suggests that post-occupancy phases of decisions are important in respect to solar energy

	Stages of Decision					
	Pre Occupancy			Post Occupancy		
Problem Aspects	Planning	Design	Modelling	Facility Use	Management	Adaptation
Solar Energy Strategy	2	3	1	1	2	0
Solar Access	3	3	2	0	1	1
Solar Geometry	2	3	2	0	0	2
Solar Energy Capture	0	3	3	2	1	1
Solar Energy Shedding	0	2	3	2	2	0
Solar Energy Storage	0	2	3	0	0	1
Solar Energy Distribution	0	2	2	2	2	0
Solar Energy Use & Control	1	2	1	3	3	1
Integration with other Systems	1	3	1	2	2	1
			Key	0	No/Little Effect	
				1	Slight Effect	
				2	Considerable Effect	
				3	High Effct	

Figure 14.3 Matrix Framework for Evaluating Passive Solar Designs and Applications
Source: Nutt, 1994

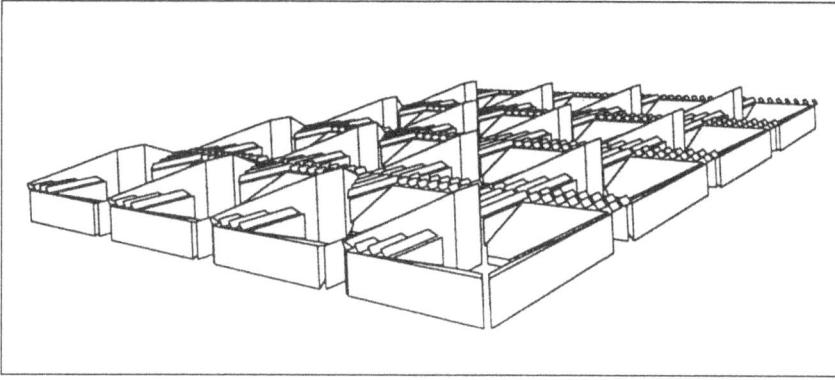

Figure 14.4 Diagram showing Photovoltaic Paneling as an Integrated Building Fabric
Source: Koester, 1994

shedding, energy distribution, energy use and control. This has been the experience with many low-energy designed buildings in Jamaica. The Petroleum Corporation of Jamaica (PCJ) Resource Centre on Trafalgar Road in Kingston is a case in point. Built according to criteria set by Dubin & Long (1978) and Dubin-Bloome (1983) as a leading-edge design for energy efficiency in 1987, its energy efficiency had been decreased by 1996 because of poor occupancy practices. This suggests that we need to be aware that passive solar design expectations may be based on optimistic assumptions about how the building will be used and be modified by those using it.

Photovoltaics can be integrated into the built environment as part of design or as independent landscape armatures. Figure 14.4 shows two differing scales at which photovoltaic technology can be integrated into urban community structures. In one way, PV panels comprise a cover for the surface of the built volume; in the other, PV panels can be treated as an independent array set outside the urban structure it supplies.

In order to increase the market impact of passive solar systems the following should be addressed:

• incentives and standards that result in improved building efficiency
• energy education for professionals that place emphasis on environmentally sensitive design
• lower priced homes that incorporate basic passive cooling techniques

In addition, sensitization is always critical and there should be more demonstration projects to educate the public, generate interest and support passive solar heating and cooling systems.

A Multifaceted Approach

Energy efficiency in buildings can be achieved through a multifaceted approach involving the use of architectural principles responsive to a number of factors. These include the climate of a particular location, the reduction of transportation energy, use of materials with low embodied energy, adoption of efficient structural design, incorporation of energy-efficient building systems, and where economically possible, the utilization of renewable energy sources such as photovoltaics and solar water heaters. The aim is to use functional and aesthetic systems of thermal environmental control, utilizing the ambient conditions of the region and the site to advantage. The visual expression of passive thermal environmental control can result in innovative, successful and exciting low energy architecture.

Sustainable development, including sustainable building design, places enormous responsibilities on all citizens in order to ensure the rational use of natural resources as well as their protection and regeneration. All countries are encouraged to cultivate a Sustainable Development Planning Framework to provide a structured model for nationally and locally led development planning. This framework would incorporate measures to ensure accountability, transparency and efficiency. The success of these measures can only be achieved through the active participation of many different kinds of stakeholders in the planning process. Stakeholders will help to maintain standards in labelling, voluntary agreements as well as the transfer of technology and capacity building.

More authority and responsibility could also be devolved to local bodies to implement an important feature of the Sustainable Development Planning Framework, which a strong national building code. A building code will educate and mandate clients, planners, and architects on the multiple benefits of building in harmony with nature. It will reinvigorate the zeal of professional designers, builders, and planners to impart to their physical development projects a bond with the earth, far greater than any other aesthetic value.

Environmentally progressive design embodies different views of life-cycle performance. The majority of existing building environment assessment methods, implicitly or explicitly assess building performance relative to conventional practice rather than in terms of progress toward sustainability. The assessment of environmental sustainability requires understanding and capturing the interaction of processes and transformations. Assessment also necessitates having absolute measures of resource use and environmental loadings associated with buildings that can be linked with known ecological limits. Future advances in environmental assessment methods will have to explicitly extend the boundaries of analysis relating to the community or area in which a building is located – thus considering regional economic and social issues along with environmental issues. The future of sustainable architectural design requires conceptual leaps and a shift to performance measures in order to properly describe advances.

In the final analysis, the reason for architects to promote green sustainable design is to improve architectural quality. Buildings with more natural and fewer artificial inputs are better. Day-lit buildings are more pleasant than artificially lit ones. If clean air is available for ventilation it is more acceptable than that from a mechanical source. So less is more, and less is beautiful. Perhaps design elegance is embodied in simple but complete solutions.

Efficiency in building design together with appropriate urban planning can make a positive impact on the reduction of energy consumption on a global basis. The cities of the future will not only be more energy efficient but also provide some of their own food supply. We will gradually see an era when farming will also be done vertically in cities through tall elevated buildings using hyponics and other advanced Greenhouse technologies. This will negate the need for more and more arable land for agriculture and the high cost of transporting agricultural products across countries. Thus in the future sustainable development will incorporate not just smart buildings in regard to energy use but an urban infrastructure that will provide some of its own food requirement. The future for building design, urban planning and landscaping will be based primarily on achieving efficiency as fuel sources become more costly.

References

Cooke, J. (1980) "Solar Architecture for the Urban Tropics". Regional seminar and workshop on solar energy applications in the tropics.

De Herde, A., and Nihoul, A. (1994) "Overheating and Day Lighting in Commercial Buildings". *Renewable Energy* 5 (pt II) pp. 917–919.

Dubin, F.S., and Long, C.G. (1978) *Energy Conservation Standards*. New York: McGraw Hill.

Dubin-Bloome Associates (1983) "Construction Manual – Energy Conservation for Buildings in Jamaica". Prepared for the Ministry of Mining and Energy, Kingston, Jamaica.

Gadi, M., B., and Ward, I.C. (1990) "Passive Utilization of Wind and Solar Energy within a composite Domed Structure for providing Natural Ventilation in Warm Climates". Proceedings 1st World Renewable Energy Congress pp. 2147–2151.

Gallo, C. (1994) "Bioclimatic Architecture". *Renewable Energy*. 5 (pt II) pp. 1021–1027.

Givoni, B. (1994) *Passive and Low Energy Cooling of Buildings*. New York: Van Nostrand Reinhold.

Hawkes, D. (1995) "Towards the Sustainable City". *Renewable Energy* 6 (3) pp. 345–352.

Kimura, K. (1994) "Vernacular Technologies Applied to Modern Architecture". *Renewable Energy* 5 (pt. II) pp. 900–907.

Koester, R.J. (1994) "The Fundamentals of Integrating "The Commons" Application as Community Tissue or Urban Implant". *Renewable Energy* 5 (pt. II) pp. 1015–1020.

Nutt, B.B. (1994) "The Use and Management of Passive Solar Environments". *Renewable Energy* 5 (pt. II) pp. 1009–1014.

Szokolay, S.V. (2002) "Solar Heat Gain through Fenestration: a Review of Current Practice". *Architectural Science Review* 45 (3) pp. 211–217.

Taleb, A.M., and Al-Wottar, A.J.H. (1988) "Design of Windows to reduce Solar Radiation Transmittance into Buildings". *Solar and Wind Technology* 3 (5) pp. 503–515.

Talmatamar, T., Alhabobi, M., Sfaxi, Y., and Awanto, C. (1995) "Analysis of Solar Radiation for Sunlit Glass Shaded by Vertical Adjustable Flat Slats". *Renewable Energy* 6 (7) pp. 663–671.

15. | Energy Efficiency

Strategies for the Future

STRATEGIES TO ENCOURAGE ENERGY EFFICIENCY will help to moderate environmental problems as well as save energy in spite of the expected growth in the world demand for energy. The world is not running out of energy. The world faces the problem of expanding its supply of energy at low direct cost. Energy intensity – the amount of energy used per unit of activity – is the inverse of energy efficiency. The energy intensity of our activities depends on how equipment is designed, operated and maintained, how well capacity is utilized and the type of energy used. In many developing countries, energy intensity is high.

Most countries have implemented energy conservation programmes which may be summarized as follows:

- the dissemination of information on energy conservation, including labelling of appliances and training for employees[1]
- the improvement in the efficiencies of household appliances
- the introduction of new energy taxes on oil-based fuels

In Jamaica, utility-provided electricity prices rose slowly in real terms after the price of oil fell in 1985. Since then there has been less economic incentive to increase energy efficiency. There has been a rapid increase in the number of homes, cars, equipment and the electrification of commercial buildings. There is a trend towards an increase in the ratio of energy use to gross domestic product (GDP). The depreciation of the dollar and the relatively high cost of electricity are expected to cause a wave of interest in energy efficiency.

Planning Energy Use in Buildings

Energy costs for running commercial buildings are usually small relative to the cost of the property and other operating costs. However, the energy bills for a commercial property can be quite large in relation to the cost of the time that a 'building engineer' would spend in improving system performance. The presence of tenants and a diverse mix of physical and economic activities and cultural forces complicate the situation. Tenants do not always choose to save energy, although they pay their own energy bills.

An effective National Energy Code for buildings is necessary to help ensure energy efficiency. This code addresses building energy efficiency in a comprehensive fashion and requires that all new buildings meet or barely exceed minimum standards to provide a cost-effective degree of energy efficiency. The code addresses at the very least the most efficient use of lighting, ventilating, air-conditioning systems and service water heating systems.

With regard to controlling or reducing costs, most industrial firms, business and government agencies regard human resource costs as their principal concern – not energy costs. This is a costly misconception. Well-executed energy plans are needed in all company operations and public buildings. In larger organizations, the appointment of an energy coordinator would ensure that there are four basic approaches to energy efficiency: measurement, assessment, evaluation and implementation.

Measurement involves analysing the current energy usage: it is not possible to make a proper energy plan without a comprehensive understanding of how a company is using and wasting energy. Measurement provides heightened empirical judgement, which in the case of energy efficiency may provide far more accurate information than personal judgement. Measurement also provides the knowledge by which the performance of energy and fuel can be objectively judged, and shows where the scope for improvement exists. Assessment poses the question as to whether or not things should continue to be done in the same way. In industry, an assessment would consider the efficiency of boilers and furnaces, steam and condensate systems, ventilation and air-conditioning, lighting, transport, as well as electrical energy and alternative energy. Evaluation involves the process by which options are chosen and priorities assigned. In this phase, evaluators check product waste, analyse fuel choices, compare practice with theory and calculate payback periods.

Solar air conditioners are an emerging technology which is now cost effective particularly for large-scale installations

Implementation can be done only if the senior management is convinced of the need. Realistic targets are critical to the eventual success of the energy plan. These targets should be achievable, measurable, time-based, and capable of being monitored.

Once energy conservation has been integrated into an organization's operations, it will be integrated into the organizational culture. Far from being just another expense to justify in the name of environment alone, experience indicates that energy efficiency improvements provide an effective way of achieving important economic objectives. These objectives include reducing costs, improving productivity, and decreasing the impact of present industrial activity on future climate change. The organization should view this as part of its social and national responsibility.

Demand Side Management

Major uncertainties and risks faced by power utility companies have resulted in a changing business philosophy, leading to increased emphasis on demand side management (DSM) as an important element of long-

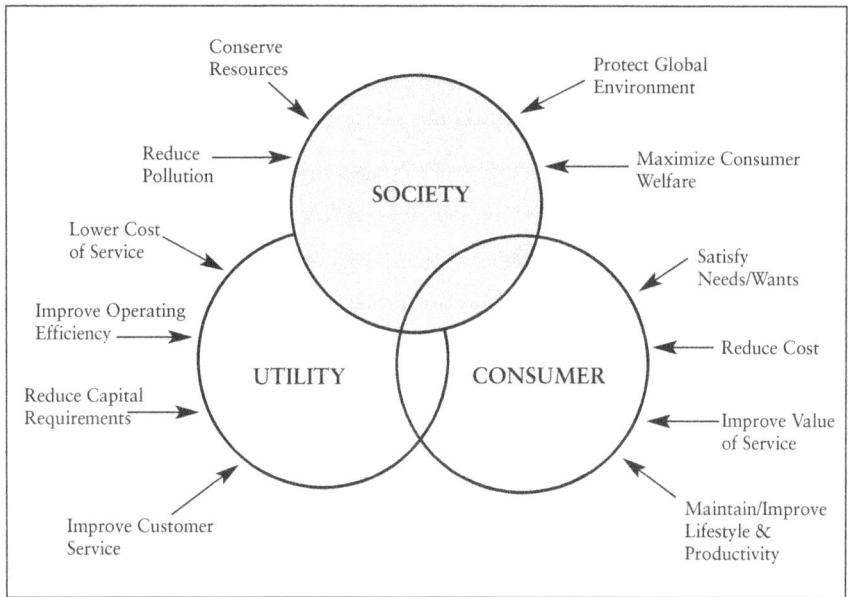

Figure 15.1 Graphic Representation of Mutually Beneficial DSM Options
Source: Limaye, 1990

term resource planning. As part of least-cost or integrated-resource planning, utility companies now approach future planning by evaluating all options, emphasizing flexibility and low risk, improving customer relations, reducing pollution and implementing least-cost growth (see Figure 15.1). The objective of DSM actions is to implement programmes that modify consumer loads with resulting benefits to the consumer, the utility company and society. The DSM process is a partnership between the customer and the utility company. The process involves among its dimensions:

- Consumer and uses
- Load-shape modification objectives
- Technology alternatives
- Market implementation methods

Consumer acceptance depends on needs, physical characteristics, fuel-use patterns, end-use load-shapes and attitudes, beliefs and perceptions about utility programmes. The load shape modification objectives include peak clipping, load shifting, valley filling, strategic conservation, and other factors that maintain a stable profile of electricity use (Limaye, 1990). The DSM implementation procedure is shown on Figure 15.2.

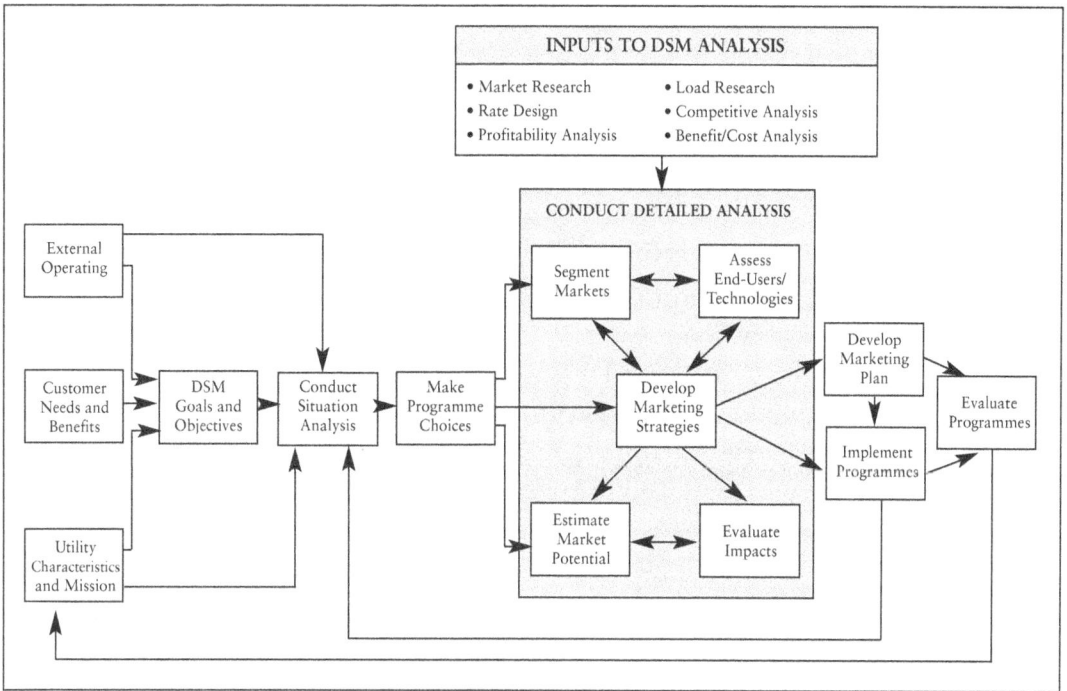

INPUTS TO DSM ANALYSIS

- Market Research
- Rate Design
- Profitability Analysis
- Load Research
- Competitive Analysis
- Benefit/Cost Analysis

CONDUCT DETAILED ANALYSIS

External Operating

Customer Needs and Benefits

Utility Characteristics and Mission

DSM Goals and Objectives

Conduct Situation Analysis

Make Programme Choices

Segment Markets

Assess End-Users/ Technologies

Develop Marketing Strategies

Estimate Market Potential

Evaluate Impacts

Develop Marketing Plan

Implement Programmes

Evaluate Programmes

Figure 15.2 Sequence in Implementation of DSM Programme

Transport

Fuel costs are low in relation to the cost of owning a motor car but high in relation to the cost of its use. Many consumers around the world are willing to trade fuel economy for power, comfort, safety, and convenience. To combat this tendency, some governments have emphasized fuel efficiency in vehicles, imposing progressively higher taxation on vehicles with larger engines. A variety of other measures to this end have been suggested, including limits on horsepower and new-car registration fees that rise with the vehicle's fuel intensity.

Studies have addressed many transport problems including air pollution, noise pollution, congestion, road adequacy, competing uses of urban space for parking and the disposal of car hulks and motor oil. These problems are difficult to solve. Possible solutions include levying additional taxes on fuel prices, imposing parking fees and charging tolls for the use of certain roads during peak hours. The implementation of proper transport policies could lead not only to more efficient vehicular traffic, but to an improved bus system, significantly reduced traffic congestion, as well as reduced air pollution.[2]

Natural Gas

Natural gas may be substituted for oil in the transport market. Natural gas can be converted to liquid, and, in the future, abundant supplies of natural gas may provide competitive transport fuel at today's prices. Natural gas is one of the cleaner fuels of the future. Gas turbine cogeneration may become a part of manufacturing since these gas turbines can be used to provide the steam needed for manufacturing as well as some electricity.

In the future, fuel cells will play an important role in commercial buildings and industrial plants. Using phosphoric acid, molten carbonate, or solid oxide electrolytes, fuel cells offer high electrical efficiencies of nearly 50 per cent. Like a battery, each cell has 'plus' and 'minus' ends separated by an electrolyte material that conducts ions just as battery acid does. Inside the cell, an electro-chemical process converts hydrogen and oxygen into electricity and water. The hydrogen is derived from natural gas, whilst the oxygen is available from the air. As long as fuel (natural gas) and air are supplied, the conversion to electrical power continues without interruption. For example, a 0.5 MW fuel cell would be able to provide a hospital with all its thermal and electrical needs and, on a co-generating basis, could make it semi-independent of the utility company and its problems with blackouts and brownouts. The hospital could sell its surplus electricity to the power company.

Future Prospects

Efficiency improvements tend to occur as nations develop, as expertise grows, as energy and other price factors are adjusted, and as people become accustomed to managing commercial fuels and electricity. The ratio of energy to output in manufacturing has fallen continually in industrialized countries. Many nations appear to suffer from the 'electrical intoxification factor', meaning that they always require more electricity.

Ultimately, efficiency improvement will occur in industry and commercial buildings owing to pressure from the utility company about peak-period power consumption, as well as rising costs. Tariffs and taxation already determine the size and power of the automobiles we drive. Future building regulations may insist that new structures incorporate energy-saving features, to the economic benefit of the occupants.

The design and placement of windows are critical to efficiency in indoor lighting

An excellent approach to energy efficiency is demonstrated by the Energy Saving Company (ESCO) in the US. ESCOs provide energy services to larger users, performing a matrix of functions. ESCOs can act as project developer, financier, and integrator of efficiency technologies. They provide rigorous measurement and verification of energy savings and often take compensation based on performance, that is, the value of energy saved. They also provide objective standards that make performance-based projects possible and also adopt sophisticated monitoring practices. Several utilities have developed performance-based financing programmes with independent ESCOs. The burden of up-front financing is placed on the ESCO which is reimbursed for its investment on a pay-for-performance basis. The utility company monitors the savings either through bill analysis or dedicated metering and pays the ESCO for the actual savings.

Efficiency is the goal of the future, and all nations should introduce efficiency measures to ensure a cleaner global environment. The public's perception of the depth and nature of environmental problems is important. Carefully developed governmental energy efficiency policies can create a market for improved efficiency services. Maintaining safeguards on the environment will over time add costs but not significantly. On the other hand, technological progress will gradually reduce costs.

The answer to environmental problems such as global warming will be found in changes in technology rather than in a dramatic reduction in energy use. An important technology that will be used more in the future is cogeneration. Cogeneration is a high-efficiency energy system that produces electricity (or mechanical power) and valuable heat from a single fuel source. Cogeneration is sometimes known as 'combined heat and power', or CHP. It offers major economic and environmental benefits because it turns otherwise wasted heat into a useful energy source. This greater efficiency means carbon dioxide emissions are cut by up to two-thirds when compared with conventional coal-fired power stations.

In essence, owners of cogeneration plants receive two major benefits, because burning fuel not only generates power, it also provides heat for industrial processes. For example, a hospital cogeneration plant could produce some of the power and all the hot water needed for its laundry needs. Similarly, office buildings can produce power for electricity and air conditioning from the waste heat generated by its air conditioning engines. Cogeneration can significantly reduce energy costs and greenhouse gas emissions, typically by up to two-thirds. Local air quality benefits can also be achieved through the replacement of older coal-fired boilers. In addition to reducing operating costs, cogeneration also increases resource utilization.

Jamaica has significant cogeneration potential using bagasse in the sugar factories. Bagasse cogeneration describes the use of fibrous sugarcane waste, bagasse, to cogenerate heat and electricity at high efficiency in sugar factories.

The benefits of bagasse cogeneration include:

- near-zero fuel costs (paid in local currency), commercial use of a waste product
- increased fuel efficiency leading to an increase in the economic viability of sugar mills
- more secure, diverse, reliable and widespread supply of electricity for local consumers
- minimal transmission and distribution (T&D) costs, and reduced network losses, as generation is located near important loads
- greater employment for local populations
- lower emissions of CO_2 and other gases than from conventional fossil-fuel generation (WADE, 2004)

The amount of energy that can be extracted from bagasse is largely dependent on two main criteria: moisture content and the technology used for energy production. The output of electricity from bagasse cogeneration plants is fundamentally dependent on the prevailing electricity market rules: inadequate buyback prices paid to mill owners by the utility company will affect the efficiency of cogeneration plants. Conversely, higher rates can provide incentive to owners to upgrade their energy facilities to enable maximum on-site efficiency. This is the key to enabling the potential for bagasse-based cogeneration to be achieved.

The Jamaican Sugar Cane industry could benefit economically from the maximal use of cogeneration technologies. It is estimated that factories such as Frome and Monymusk could produce between 40 and 50 megawatts of electricity from bagasse and that the smaller factories could produce of the order of 20 MW each. An additional benefit can be derived by the trading of carbon credits under the Clean Development Mechanism (CDM).

The CDM is an arrangement under the Kyoto Protocol allowing industrialized countries with a greenhouse gas reduction commitment (referred to as Annex 1 countries) to invest in projects that reduce emissions in developing countries as an alternative to more expensive emission reductions in their own countries. To receive credit, emissions reductions should be additional to what would occur in the absence of the CDM (Meyers, 1999). As of July 21, 2008, the CDM Executive Board has registered 1,128 projects as CDM projects. These projects reduce greenhouse gas emissions by an estimated 220 million tonne CO_2 equivalent per year.[3]

Every high school student should understand the reason for CDM, and what can be done to ensure that the CDM works. In order to change the national perception of energy efficiency, we have to concentrate on the younger generation and inculcate in students a sound attitude towards energy use. The negative effect of poor energy efficiency over the past 200 years suggests that the next generations will deal with the consequences. There is therefore a certain urgency to teach students the tenets of energy efficiency from a young age. This may well be the strongest power and force of the next generation.

References

Limaye, D.R. (1990) "Demand-Side Management: Developing Least-Cost Strategies to Reduce Electricity Costs and Protect the Global Environment". International Conference on Energy and Environment, Bangkok, Thailand, November 27–30, pp. 24–33.

Meyers, S. (1999) "Additionality of Emissions Reductions from Clean Development Mechanism Projects: Issues and Options for Project-Level Assessment" Ernest Orlando Lawrence Berkeley National Laboratory. Retrieved from: http://ies.lbl.gov/iespubs/43704.pdf

World Alliance for Decentralized Energy (WADE) (2004), "Bagasse Cogeneration – Global Review and Potential", June.

16. | The Hydrogen Economy

A New Energy Agenda

HYDROGEN COULD BECOME IMPORTANT AS a primary fuel after 2030. In the meantime, natural gas is being considered as the transition fuel towards a possible hydrogen economy. The fuel cell is the technology that will use hydrogen to produce energy. A fuel cell is an electrochemical energy conversion device. It produces electricity from the various external quantities of fuel (on the anode side) and an oxidant (on the cathode side): these react in the presence of an electrolyte. Generally, the reactants flow in and reaction products flow out while the electrolyte remains in the cell. Fuel cells can operate virtually continuously as long as the necessary flows are maintained.

Fuel cells are different from batteries in that they consume reactant, which must be replenished, whereas batteries store electrical energy chemically in a closed system. Additionally, while the electrodes within a battery react and change as a battery is charged or discharged, a fuel cell's electrodes are catalytic and relatively stable: therefore, unlike a battery, the fuel cell does not run down provided it is fed with hydrogen.

Many combinations of fuel and oxidant are possible. A hydrogen cell uses hydrogen as fuel and oxygen as oxidant. Other fuels include hydrocarbons and alcohols, and other oxidants include air, chlorine and chlorine dioxide. The efficiency of a fuel cell is dependent on the amount of power drawn from it. Drawing more power means drawing more current, which increase the losses in the fuel cell. As a general rule, the more power (current) drawn, the lower the efficiency. Most losses manifest themselves as a voltage drop in the cell, so the efficiency of a cell is almost proportional to its voltage.

Types of Fuel Cells

Fuel cells are classified primarily by the kind of electrolyte they employ: this determines the kind of chemical reactions that take place in the cell, the kind of catalysts required, the temperature range in which the cell operates, the fuel required and other factors. These characteristics in turn affect the applications for which these cells are most suitable. There are several types of fuel cells currently under development, each with its own advantages, limitations and potential applications. Some of the most promising types include:

- Polymer Electrolyte Membrane (PEM)
- Phosphoric Acid
- Direct Methanol
- Alkaline
- Molten Carbonate
- Solid Oxide

Polymer Electrolyte Membrane

Polymer electrolyte membrane (PEM) fuel cells, also referred to as proton exchange membrane fuel cells, deliver high power density and offer the advantages of low weight and volume, compared to other fuel cells. PEM fuel cells use a solid polymer as an electrolyte and porous carbon electrodes containing a platinum catalyst. They need only hydrogen, oxygen from the air, and water to operate and do not require corrosive fluids like some fuel cells. They are typically fueled with pure hydrogen supplied from storage tanks or onboard reformers.

PEM fuel cells operate at relatively low temperatures, around 80°C (176°F). Low temperature operation allows them to start quickly and results in less wear on system components, resulting in better durability. A noble-metal catalyst (typically platinum) should be used to separate the hydrogen's electrons and protons, which adds to system costs. The platinum catalyst is also extremely sensitive to carbon monoxide (CO) poisoning, making it necessary to employ an additional reactor to reduce CO in the fuel gas if the hydrogen is derived from an alcohol or hydrocarbon fuel. This also adds to the cost. Developers are currently exploring platinum/ruthenium catalysts that are more resistant to CO.

PEM fuel cells are used primarily for transportation applications and

some stationary applications. Due to their fast startup time, low sensitivity to orientation and favourable power-to-weight ratio, PEM fuel cells are particularly suitable for use in passenger vehicles such as cars and buses. A significant barrier to using these fuel cells in vehicles is hydrogen storage. Most fuel cell vehicles (FCVs) powered by pure hydrogen must store the hydrogen onboard as a compressed gas in pressurized tanks. Due to the low energy density of hydrogen, it is difficult to store enough hydrogen onboard to allow vehicles to travel the same distance as gasoline-powered vehicles before refueling, typically 300–400 miles. Higher-density liquid fuels such as methanol, ethanol, natural gas, liquefied petroleum gas and gasoline can be used for fuel, but the vehicles must have an onboard fuel processor to reform the methanol to hydrogen. This increases costs and maintenance requirements. The reformer also releases carbon dioxide, though less than that emitted from current gasoline-powered engines.

Phosphoric Acid

Phosphoric acid fuel cells (PAFCs) use liquid phosphoric acid as an electrolyte. This type of fuel cell is typically used for stationary power generation, but some PAFCs have been used to power large vehicles such as city buses. PAFCs are more tolerant of impurities in the reformate than PEM cells, which are easily "poisoned" by carbon monoxide. Carbon monoxide binds to the platinum catalyst at the anode, decreasing the fuel cell's efficiency. PAFCs are 85 per cent efficient when used for the co-generation of electricity and heat, but less efficient at generating electricity alone (37 to 42 per cent): this in itself is only slightly more efficient than combustion-based power plants, which typically operate at 33 to 35 per cent efficiency. PAFCs are also less powerful than other fuel cells, given the same weight and volume. As a result, these fuel cells are typically large and heavy. PAFCs are also expensive. Like PEM fuel cells, PAFCs require an expensive platinum catalyst, which raises the cost of the fuel cell. A typical phosphoric acid fuel cell costs between US$4,000 and US$4,500 per kilowatt to operate.

Direct Methanol

Most fuel cells are powered by hydrogen, which can be fed to the fuel cell system directly or can be generated within the fuel cell system by reforming hydrogen-rich fuels such as methanol, ethanol, and hydrocarbon fuels. Direct methanol fuel cells (DMFCs), however, are powered by pure methanol, which is mixed with steam and fed directly to the fuel cell anode. Direct methanol fuel cells do not have many of the fuel storage problems typical of some fuel cells since methanol has a higher energy density than hydrogen, though less than gasoline or diesel fuel. Methanol is also easier to transport and supply to the public as it is a liquid, like gasoline. Direct methanol fuel cell technology is relatively new compared to that of fuel cells powered by pure hydrogen, and research and development are roughly three to four years behind that of other fuel cell types.

Alkaline

Alkaline fuel cells (AFCs) were one of the first fuel cell technologies developed, and were the first type widely used in the U.S. Space Program to produce electrical energy and water onboard spacecraft. These fuel cells use a solution of potassium hydroxide in water as the electrolyte and can use a variety of non-precious metals as a catalyst at the anode and cathode. High-temperature AFCs operate at temperatures between 100°C and 250°C (212°F and 482°F). More recent AFC designs operate at lower temperatures of roughly 23°C to 70°C (74°F to 158°F).

AFCs are high-performance fuel cells due to the rate at which chemical reactions take place in the cell. They reach efficiencies of 60 per cent in space applications. The disadvantage of this fuel cell type is that it is easily poisoned by carbon dioxide (CO_2). In fact, even the small amount of CO_2 in the air can affect the cell's operation, making it necessary to purify both the hydrogen and oxygen used in the cell. This purification process is costly. The cell's susceptibility to poisoning also affects its lifetime (the amount of time before it must be replaced), further adding to cost.

Cost is less of a factor for remote locations such as space or under the sea. To effectively compete in most mainstream commercial markets, these fuel cells will have to become more cost effective. AFC stacks have been shown to maintain sufficiently stable operation for more than 8,000 operating hours. To be economically viable in large-scale utility applications, these fuel cells need to reach operating times exceeding 40,000 hours. This is possibly the most significant obstacle in commercializing this fuel cell technology.

Molten Carbonate

Molten carbonate fuel cells (MCFCs) are currently being developed for natural gas and coal-based power plants for electrical utility, industrial, and military applications. MCFCs are high-temperature fuel cells that use an electrolyte composed of a molten carbonate salt mixture suspended in a porous, chemically inert ceramic lithium aluminum oxide (LiA_1O_2) matrix. Since they operate at extremely high temperatures of 650°C (approximately 1,200°F) and above, non-precious metals can be used as catalysts at the anode and cathode, thus reducing costs.

Improved efficiency is another reason MCFCs offer significant cost reductions over phosphoric acid fuel cells (PAFCs). Molten carbonate fuel cells can reach efficiencies approaching 60 per cent, considerably higher than the 37–42 per cent efficiencies of a phosphoric acid fuel cell plant. When the waste heat is captured and used, overall fuel efficiencies can be as high as 85 per cent. Unlike alkaline, phosphoric acid and polymer electrolyte membrane fuel cells, MCFCs do not require an external reformer to convert more energy-dense fuels to hydrogen. Due to the high temperatures at which they operate, these fuels are converted to hydrogen within the fuel cell itself by a process called internal reforming, which also reduces costs.

Molten carbonate fuel cells are not prone to carbon monoxide or carbon dioxide "poisoning" – they can even use carbon oxides as fuel – making them more attractive for fueling with gases made from coal. Although they are more resistant to impurities than other fuel cell types, scientists are looking for ways to make MCFCs resistant enough to impurities from coal, such as sulfur and particulates.

The primary disadvantage of current MCFC technology is durability. The high temperatures at which these cells operate and the corrosive electrolyte used accelerate component breakdown and corrosion, decreasing cell life. Scientists are currently exploring corrosion-resistant materials for components as well as fuel cell designs that increase cell life without decreasing performance.

Solid Oxide

Solid oxide fuel cells (SOFCs) use a hard, non-porous ceramic compound as the electrolyte. Since the electrolyte is a solid, the cells do not have to be constructed in the plate-like configuration typical of other fuel cell types. SOFCs are expected to be around 50–60 per cent efficient at converting fuel to electricity. In applications designed to capture and utilize the system's waste heat (cogeneration), overall fuel use efficiencies could top 80–85 per cent.

SOFCs operate at very high temperatures – around 1,000°C (1,830°F). High temperature operation removes the need for precious-metal catalysts, thereby reducing cost. It also allows SOFCs to reform fuels internally, which enables the use of a variety of fuels and reduces the cost associated with adding a reformer to the system. SOFCs are also the most sulfur-resistant fuel cell type; they can tolerate several orders of magnitude more sulfur than other cell types. In addition, they are not poisoned by CO, and can even be used as fuel. This allows SOFCs to use gases made from coal.

The high-temperature operation of SOFCs has disadvantages. It results in a slow startup and requires significant thermal shielding to retain heat

Fuel cells are now being used in the transport sector, particularly for urban buses

and protect personnel, which may be acceptable for utility applications but not for transportation and small portable applications. The high operating temperatures place stringent durability requirements on materials. The development of low-cost materials with high durability at cell operating temperatures is the key technical challenge facing this technology.

Scientists are currently exploring the potential for developing lower-temperature SOFCs operating at or below 800°C; these will have fewer durability problems and cost less. Lower-temperature SOFCs produce less electrical power, and stack materials that will function in this lower temperature range have not been identified.

Regenerative (Reversible) Fuel Cells

Regenerative fuel cells produce electricity from hydrogen and oxygen and generate heat and water as by-products, just like other fuel cells. Regenerative fuel cell systems can also use electricity from solar power or some other source to divide the excess water into oxygen and hydrogen fuel – this process is called "electrolysis." This is a comparatively young fuel cell technology being developed by NASA and others.

The Ultimate Fuel for the 21st Century

The current interest in the hydrogen economy derives from the fact that in assessing future fuel technologies hydrogen is regarded as the ultimate "fuel" for the twenty first century. This will obtain provided the hydrogen is derived from renewable sources. Hydrogen has near-zero emissions of both local pollutants and GHGs when used with fuel cells (Owen, 2005: 71). There are giant hydrogen gas grids around the world and hydrogen is also being tanked around in liquid form in ships: according to Elliott (2003), this is already being done with hydrogen from Canada's hydro plants being shipped to Sweden. In the longer term, countries in the Middle East might install large arrays of photovoltaic solar cells in desert areas to generate hydrogen to supply the less sunny areas of the world with a new clean fuel. As illustrated in this chapter, hydrogen can be used as a clean fuel for vehicles or can be converted into electricity in a fuel cell. Fuel cells are also increasingly being used for stationary power generation and by 2012 there could be 16,000 MW of stationary fuel cell capacity in use. In the future there may be a global energy system based on renewable hydrogen, with electricity only being generated where and when required.

References

Elliott, D. (2003) *Energy, Society & Environment*. New York: Routledge.
Owen, A.D. (2005) "Burning Up: Energy Usage and the Environment".
   Harvard International Review 26 (4) pp. 66–71

17. | Future Energy Supply Options

Diversifying and Expanding the Energy Mix

A NEW ENERGY AGENDA WILL change the present energy mix. Comparative scarcity due to political or resource constraints will increase the relative price and availability of oil. The cost of reducing environmental degradation will affect the current cost competition among fuels, and the fear of global climate change will encourage a shift from carbon-based fuels to nonfossil alternatives. An energy resource perspective for the next fifty years involves speculation on future energy demand, probable economic energy supply alternatives, and possible changes in energy systems and technologies. The expectations of the future are shaped and directly affected by the long periods required to develop new or improved energy technologies and to install them commercially. Long-term projections generally rely on judgement based on present knowledge and experience, and the recognition of trends, and so help to guide present plans, projects and strategies.

In the context of the time scale of the ensuing decade, an obvious question is whether or not the cumulative effect of the gradual depletion of carbon-based energy resources will result in a global limitation on energy systems, especially on the future supply of liquid fuel for transportation purposes. Present technology points to the potentially large-scale convertibility of all fossil fuels to liquid forms or gas and therefore a global source of liquid fuel could be derived from the large known coal reserves. The rising cost of oil will bring into competition known higher-cost sources such as coal conversion, tar sands, and oil shale.

Table 17.1 shows the world's marketing energy consumption. Although the world demand for primary energy is increasing with the

Table 17.1   World Marketed Energy Consumption by Country Grouping (Quadrillion Btu)

Region	2005	2010	2015	2020	2025	2030	Average Annual Per cent Change 2005–2030
OECD	240.9	249.7	260.5	269	277.6	285.9	0.7
North America	121.3	126.4	132.3	137.8	143.4	148.9	0.8
Europe	81.4	83.9	86.8	88.5	90.4	92.0	0.5
Asia	38.2	39.3	41.4	42.7	43.7	44.9	0.7
Non-OECD	221.3	262.8	302.5	339.4	374.2	408.8	2.5
Europe and Eurasia	50.7	55.1	59.5	63.3	66.0	69.1	1.2
Asia	109.9	137.1	164.2	189.4	215.3	240.8	3.2
Middle East	22.9	26.4	29.5	32.6	34.7	36.8	1.9
Africa	14.4	16.5	18.9	20.9	22.5	23.9	2.0
Central and South America	23.4	27.7	30.5	33.2	35.7	38.3	2.0
Total World	462.2	512.5	563.0	608.4	651.8	694.7	1.6

Note: Totals may not equal sum of components due to independent rounding

Sources: 2005: Energy Information Administration (EIA), International Energy Annual 2005 (June-October 2007) website www.eia.doe.gov/iea, Projections: EIA World Energy Projections Plus (2008)

highest energy consumption in China and the industrialized countries, it is unlikely that a serious shortage of fuels will develop worldwide during this century, even with the depletion of coal, oil, and gas resources. Yet fossil fuel reserves are ultimately finite. The main uncertainties arise from the constantly changing economic competition among the various fuel sources and the effects of environmental factors which incur increased costs because of the need to reduce the undesirable effluents from fuel use. A fact to be borne in mind is that the real resource cost of energy is lower today than at the start of this century, while the world's population has tripled and its economic output increased enormously.

Figure 17.1 shows the primary energy consumption by fuel type indicating that oil is still the major fuel source. Other fuels such as coal, nuclear and biomass will continue to rise in relative importance. There

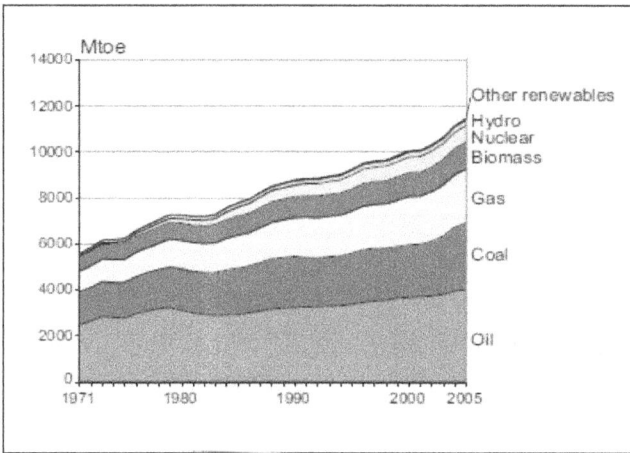

Figure 17.1 World Primary Energy Consumption by Fuel Type
Source: Sims et al 2007

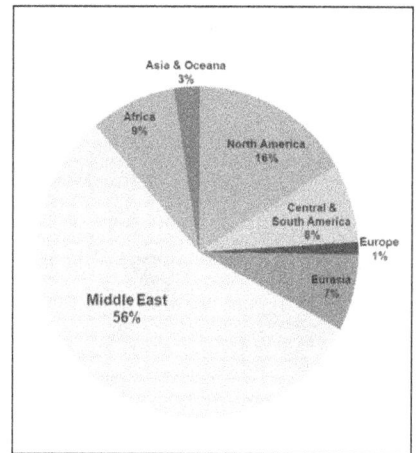

Figure 17.2 World Oil Reserves by Region

are approximately one trillion barrels of proven oil reserves, enough to last until 2050 at the current rates of consumption. Figure 17.2 shows the world reserves by region, indicating that much of the oil reserve lies in the Middle East.

Over time, technology and economic incentives have overcome apparent limitations of resources. Nonetheless, it is likely that real costs of primary energy from conventional oil and gas sources will increase because the average cost of exploration and development of new oil fields has risen steadily. At some point in the increase of the price for conventional oil and gas sources, unconventional oil sources will become competitive. Unconventional sources will however require large capital investments. For example, at an oil price of US$35 per barrel, large high-cost oil resources such as tar sands become economically viable. The reserves of oil shales and tar sands are sufficient to provide the liquid fuel requirements for much of the remainder of this century but they need much higher capital investment for the same flow rates. If these are utilized, the character of the liquid fuel production industry will be changed.

Coal is the most abundant fossil fuel, representing nearly 80 per cent of all known conventional fossil resources. It can be converted into both liquid and gaseous hydrocarbons. The production of complex hydrocarbons by coal conversion has been deployed commercially worldwide. All such coal conversion plants burn coal in the presence of an oxidant and water vapour, leading to the decomposition of the steam and reaction with the hot coal to produce syngas (carbon monoxide and hydrogen) –

the ratio of each component depending on the system parameters. Subsequent treatment can be applied to produce a spectrum of hydrocarbon mixtures, including hydrogen. The production of methane or liquid fuel thereafter is well-established engineering. Since 1985, a plant in New Zealand has demonstrated the feasibility of converting methane into gasoline with a zeolite process. The Sasol plant in South Africa has been producing liquid fuel from indigenous coal for more than two decades. The true costs are competitive with oil at US$25 per barrel.

Large natural gas reserves in some remote areas have encouraged the investment in on-site conversion of natural gas to liquid fuel to facilitate the shipment of the product to distant markets. It is obvious that if the price of oil becomes sufficiently high, the conversion of coal to liquid fuel will enter the energy industry investment portfolio.

The use of coal conversion to gas for generating electricity has been demonstrated by the 100 MW integrated gasification combined cycle (IGCC) power plant of Southern California Edison built in the late 1980s. It is the cleanest coal-fuelled technique available. A more advanced integrated gasification humid air turbine cycle (IGHAT) has been proposed. This eliminates the steam bottoming cycle by using the residual heat in a modified combustion turbine. These advanced cycles are important steps for improving the efficiency of coal use. At this time, a modern commercial coal station has an efficiency of about 34 per cent. The coal conversion IGCC has an efficiency of 38 per cent, and the IGHAT has an efficiency of approximately 41 per cent. The advanced cycles have a higher capital cost, but with continuing development they may become competitive, especially where environmental considerations are important.

A significant improvement in the efficiency of fossil fuel generation will arise from the development of the fuel cell. The molten carbonate fuel cell is the present focus of developmental work. An electrochemical process is used to move from hydrogen and oxygen in gas to electricity. The hydrogen is derived from natural gas by a process called reforming, while the oxygen is available from the air. As long as gas and air are supplied, conversion to electricity continues without interruption. Thus fuel cells need no recharging. In principle the fuel cell directly replaces the combustion turbines in the integrated cycles described above. It would raise the efficiency of the two cycles to 46 and 55 per cent respectively. The commercialization of the fuel cell would eventually decrease the electricity component of the global demand for coal to approximately 65 per cent of that needed in present plants, thus reducing annual carbon emissions.

The importance of renewables and nonfossil fuel sources in future global energy use is directly dependent on their economic competitiveness. This group includes solar, wind, biomass, geothermal, hydropower and nuclear. Only hydropower and nuclear are important contributors today, with hydropower providing about 20 per cent of global electricity and nuclear about 16 per cent. There are limitations on the potential contributions of the renewable and nonfossil sources. The energy input needed to manufacture the renewables and their high initial capital costs are major issues. The question of net energy output, that is, output less the energy input from other resources used for their manufacture is especially pertinent to biomass, where the energy input for their growth (for instance fertilizer and water) and processing are substantial. It is yet to be determined over the lifetime of the unit if the cost of energy delivered is high.

The economics of hydropower and nuclear power are the best understood of all the renewables. They both need approximately the same capital investment per plant (per MW), about twice that of a coal-fired unit. When compared to coal, hydropower has no fuel cost and low maintenance and operating costs. Nuclear has a low fuel cost and high maintenance and operating costs. In the highly industrialized countries, nuclear power is now generally competitive with coal. The global growth of hydropower could be increased by four times the current level before reaching the upper limit. However, both technologies are constrained by environmental issues. Hydropower expansion requires flooding large areas and changing river flows. Nuclear energy has a small environmental impact but is restrained by public concerns about the risk of accidental release of radioactivity from either the reactor or used fuel, hence nuclear has high potential social penalties. For this reason, engineering developments have concentrated on ways to reduce the likelihood of accidents. When assessing comparative risks to public health, safety and environment, nuclear is a better choice than coal.

The economics of bioenergy is more difficult to analyse in the developing world, because the sources are often non-commercial (wood, agricultural wastes, animal dung) and in some cases require no capital, only labour. The true cost of the non-commercial sources in the developing countries is speculative because it is not market priced. The labour cost to collect such non-commercial sources is high. As the economies of developing countries evolve, there will be a shift from labour to less expensive productive activity with a gradual move from non-commercial to commercial fuels, which initially could be petroleum products (Starr et al., 1992).

The use of commercial biomass fuel production through managed agriculture and forestry has been studied. Ethanol production from sugar cane in Brazil is a demonstration of commercial biomass utilization. Ethanol from sugar cane has a significant higher energy output ratio than ethanol produced from corn: the energy input/output ratio for sugar cane ethanol in Brazil is 1:8 whereas the input/output energy ratio for corn ethanol produced in the USA is no more than 1:3. The tropics are the optimal areas for production of biomass through managed forestry. Also, managed forestry can have a positive environmental impact. As the reduction of carbon emissions become a global priority, managed biomass production is highly weighted because it sequesters atmospheric carbon or recirculates it. Perhaps managed forestry should be used to sequester carbon rather than use it as a fuel, because of its uncertain net energy contribution.

Solar and wind sources, although of low efficiency, have the potential to provide a significant proportion of future world energy needs. The major difficulty is overcoming their intermittent nature with energy storage and expanded collectors. Unless a low-cost electricity storage device is developed, the large-scale participation of solar and wind sources will be limited to about 10 per cent of the network capacity of fossil-based electricity systems.

It has been suggested that solar electricity should be used to produce hydrogen from water, to be used as a transportation fuel. This would achieve an ideal system of energy storage: no carbon emissions or pollutants and an external primary energy resource. Although scientifically sound, there are practical barriers of economics and possible technological problems. Many billions of cubic feet of pure hydrogen are produced by the world's oil refineries at a fraction of the cost of electrolytic hydrogen. Nonetheless, there is no indication that a transition from conventional energy systems to hydrogen-based systems has been of practical interest. There is nothing to suggest that economic and technical constraints will be removed, although hydrogen combustion producing only water as a by-product is a tantalizing concept with obvious merits.

The history of changes in fuel patterns (wood-coal-oil) suggests that in a peacetime commercial environment four or five decades are needed to change the patterns of resource use. Four advanced sources fill the horizon:

1. IGCC coal conversion based systems[1]
2. Natural gas-based liquid fuels and chemicals
3. Direct production of liquid fuels from coal (transportation fuels)
4. Molten carbonate fuel cell electricity

Molten carbonate fuel cell technology, in particular, has just emerged from the research and development phase and small 200-kW sized units have been field tested. The plan is for small units, in the early stage of this technology, to operate on natural gas or distillate fuels and to be connected to the electricity grid system. Large systems (50–250 MW), based on synthesis gas from coal gasification systems, still require further development. A feature of such plants is high efficiency (55 per cent), low emissions of carbon dioxide and no production of nitrogen oxides or sulphur. Fifty years may be required to install 100,000 MW of fuel cell equipment. To emphasize the long lead times required, we should recognize that catalytic cracking, which became commercially available in 1942, took approximately two decades to be generally used in oil refineries.

The long lead time required for fuel source transitions has important implications for global energy strategies. Renewable energy also fits into this time scale of 30 to 50 years. By 2050, about 30 per cent of the world's electricity generating capacity might be advanced technology. In developed countries, the installation of advanced energy technologies is limited by the slow obsolescence of existing plants and by the pace at which additional capacity is required. With a long-term growth rate of two per cent per annum, approximately 36 years is needed to double total capacity. In the case of the developing countries, the scarcity of capital and avoidance of safety in performance are more important than obsolescence. For this reason developing countries are more interested in installing well-proven, reliable, conventional plants. Developing countries usually require small and dispersed power growth, providing a particular niche for solar and wind, as well as small conventional carbon-based units.

Energy conservation is often regarded as being equivalent to an energy resource because it extends the life of conventional resources.

There are two main approaches:

1. Changing end-user habits to reduce energy consumption

*Alternative fuels
including biofuels
are in an incipient
state of development
for aviation*

Alternative fuels including biofuels are in an incipient state of development for aviation

2. Improvements of efficiencies in equipment and appliances

Using public transportation rather than individual automobiles is an example of the first approach. Both strategies depend on cost savings. In the 1970s, US industry invested in equipment to reduce energy use and the subsequent economic effect was positive. On the other hand, the non-industrial end-user is less influenced by long-term economies and usually waits for the normal obsolescence of equipment. Therefore, the time for a shift to more efficient energy use is between 15 and 30 years in developed countries, depending on the incentives.

In modifying consumer habits to reduce energy demand, the cost of energy relative to consumer income should be considered: this is affected by the priorities of the consumer for various services, convenience, comfort, and time-saving. For example, the driver of an automobile may find the time saved by travelling rapidly to be more valuable than the additional cost of the fuel. The tendency will be to support low cost energy, regardless of the intangible social costs of environmental damage. Without strong economic incentives, the implementation of pervasive global conservation programmes can be as long as the lead time for advance technologies, that is, approximately 50 years.

Economic growth that is compatible with reducing pollution can in some measure be accomplished by improving energy efficiency. The positive effects of energy efficiency are limited only by the additional capital that may be required to implement the most efficient systems. If the reduction of carbon emissions were to become a worldwide objective, a move to non-carbon fuel sources would be accelerated. Perhaps a commercially viable fuel cell will offer a much-used means of converting gasified fossil fuels to electricity, thereby reducing polluting emissions by about 30 per cent of today's levels.

The Future

In the future, most energy sectors will experience greater growth. Where global hydropower resources are concerned, only approximately 25 per cent of these resources have been developed and more growth is expected in this sector, although there are ecological limits. The development of simpler, safer nuclear power plants may encourage an increase in the use of nuclear power as well as small pebble bed reactors: it should be borne in mind that nearly 75 per cent of France's energy demand is satisfied from nuclear plants. Solar and windpower will play an increasing role in the development of an efficient electricity storage system if the difficulties of interrupted supply are overcome.

With regard to end-use systems, there are evolving technologies that can reduce oil use and its adverse environmental impact. Electric automobiles may become marginally competitive in some special situations, such as heavily polluted urban areas. Magnetically levitated versions of electric trains may benefit from new superconductivity developments. More efficient space heating and cooling equipment such as advanced heat pumps may move into common use.

The future we face involves an array of flexible strategies in energy supply, energy efficiency, environmental sustainability, and economic growth. A balanced programme of energy development offers the best prospects of achieving the lowest social and environmental costs. What people demand is not energy but the services which energy can provide – cooling, heating, lighting, cooking and transportation. In the future, the focus will be on how to provide these services most effectively and efficiently with minimum pollution – eventually by means different from those which are now familiar.

Future investments in Technology Research, Development, Demonstration, plus Deployment will, in part, determine:

- the future security of energy supplies
- the availability and affordability of energy services
- the attainment of sustainable growth
- the free-market distribution of energy supplies to all countries
- the deployment of low-carbon energy carriers and conversion technologies; the quantities of GHGs emitted for the rest of this century
- the achievement, or otherwise, of GHG stabilization concentration levels (Sims et al., 2007)

Technological developments will foster the reduction of global energy intensity. In addition to new and improved energy-conversion technologies, such concepts as novel supply structures, distributed energy systems, grid optimization techniques, energy transport and storage methods, load management, cogeneration and community-based services will have to be upgraded. It will require an expansion of the worldwide knowledge base on energy systems and in technologies. Major innovations that will shape society will require a foundation of strong basic research. In addition, more interdisciplinary research is required in the geosciences, material science and engineering in order to bring about new and better energy services.

References

Sims, R.E.H., Schock, R.N., Adegbululgbe, A., Fenhann, J., Konstantinaviciute, I., Moomaw, W., Nimir, H.B., Schlamadinger, B., Torres-Martínez, J., Turner, C., Uchiyama, Y., Vuori, S.J.V., Wamukonya, N., Zhang, X. (2007) "Energy supply". In *Climate Change 2007: Mitigation*. Contribution of Working Group III to the Fourth Assessment Report of the Intergovernmental Panel on Climate Change, Metz, B., Davidson, O.R., Bosch, P.R. Dave, R., Meyer L.A. (Eds). New York: Cambridge University Press.

Starr, C., Searl, M.L., Alpert, S. (1992) "Energy Sources: A Realistic Outlook". *Science*, 25b (981) pp. 9–18.

18. | Sustainable Energy

A Renewables Manifesto

> The most effective way to minimize the potentially overwhelming capital and environment cost associated with energy development is to adopt a comprehensive national energy policy that integrates both energy supply and energy-efficiency options into an overall strategy to meet energy service needs, opting first for those options with the lowest capital and environmental costs.
>
> – Michael Phillips, 1991

WORLD ENERGY CONSUMPTION HAS INCREASED by over 300 per cent in the last fifty years. The greatest level of increase has occurred in the developing countries led by China, Brazil and India. In the future, each country will be on the quest for a secure supply of energy at the lowest economic and social cost. There will be a gradual increase in oil prices over time as the resources are being consumed without significant addition to reserves.

The burning of fossil fuels is adding CO_2 into the biosphere (atmosphere/ocean/biomass) at rates that may meet or exceed any that have occurred in Earth's history. About half of the carbon emitted has been taken up by the ocean[1] and terrestrial biota. Projections into the future demonstrate that the ability of these sinks to accommodate fossil fuel CO_2 will be reduced, leading to an increase in the airborne fraction of CO_2 (Kump et al., 2009). Long-term cooperative action is needed to combat climate change, in order to stabilize greenhouse gas concentrations in the atmosphere at a level that would prevent dangerous anthropogenic interference with the climate system.

Energy Efficiency

Energy efficiency not only serves to help mitigate climate change, it also helps the economy of most nations. The high cost of the energy resources imported into most developing nations is encouraging the adoption of measures to improve energy efficiency and conservation. Fortunately, energy efficiency is now popularly promoted in many countries as an important response to environmental and economic concerns. At present there is inefficient use of energy, particularly in households and office buildings. Energy efficiency will form a major element of strategies to meet the new commitments under the Climate Convention.

The imposition of taxes is one energy option for enhancing energy efficiency, but attempts to introduce energy taxes in the European Union and the United States have not met with positive public reaction. The interests opposing higher energy prices through taxation are formidable and in many countries the burden of taxation is being shifted from income to consumption. Taxes on carbon-based nonrenewable energy would be a natural part of this reform, with tax benefits given to the use of renewable energy resources. Demand Side Management (DSM) is a programme being driven in a number of countries, and over time the programme is expected to result in reductions in per capita consumption.

The matters addressed in DSM range from lighting to integrated management systems, from simple lamp replacement to entirely new lighting systems and from motor replacements to variable speed drives. One of the first steps is to establish the technical potential for savings, of both capacity and energy, for each technology or approach. There is also the need to reduce the technical potential in line with the market potential, because not every kilowatt of energy will be saved through implementation of available technologies. Usually market potential ranges between 30 and 60 per cent of technical potential.[2]

We need a sustainable energy system in which inputs and outputs are compatible with nature and do not upset its equilibrium. It is difficult to conceive of such a system, but perhaps the wisdom is to recognize that there are no final solutions to the problem of sustainable energy growth. Among the solutions to encourage energy efficiencies are policies and financial incentives.

Policy and Financial Incentives

Public policy should not only be adaptive but anticipatory. Public policy should consider the fact that national energy systems do not respond quickly to change because of intractable infrastructure (thermal power plants may last 30 to 40 years), and long lead times (10 to 15 years) from concept to commissioning. Because of this, we should evaluate carefully our choices for industrial fuels in the future. The emphasis should be placed on clean fuels and indigenous alternative energy sources. Energy and environmental goals should work in tandem with each other so that sustainable energy policy will:

– improve the efficiency of energy use
– reduce the environmental impacts of economic activities
– minimize the depletion of finite energy resources
– reduce waste and pollution
– encourage the use of renewable energy
– enhance the diversity of the energy mix

A number of issues are to be considered in developing public policies and the legislation and regulation to implement these policies. Among the matters to be considered are:

• energy efficient building codes
• economic incentives for using environmentally friendly technologies and reducing pollutants
• fiscal incentives for low energy and environmentally benign construction in keeping with building codes
• a nominal tax on all electric utility bills with revenue dedicated to funding renewable resources and efficient technology programmes
• mandatory least-cost utility planning
• an unweighted balance in setting rates for supply-side and demand-side resources
• prudency review criteria for renewable resources investments
• cost recovery mechanisms for DSM expenditures

Against this background, utility companies can develop an integrated resource plan with:

– reasonably reliable short and long-term load forecasts
– projected costs for construction of new utility-owned generation

- financial forecasts including rates of inflation and interest rates on loan capital
- projected generation costs for fuel, operating, and maintenance expenses
- sources, costs and reliability of renewable resources
- availability, costs, reliability, load impacts and market potential of energy efficient technologies

Once resource needs are established, it is possible to determine which renewable strategies and DSM activities are optimal to the implementation of a least-cost plan. Utility companies can use many approaches to help the customer afford DSM in the economic market. Such approaches include loans, shared savings plans, leasing and rate incentives as a means to encourage changes in customer usage. A non-market driven mechanism to support renewable energy projects is the sale of government obligation bonds to finance small-scale energy projects. Such bonds can be specific to energy saving in public-owned facilities, and to private projects that use renewable resources whether for business or home use.

A number of countries have used creative ways to foster renewable energy power projects. In 1989, the United Kingdom developed the Non-fossil Fuel Obligation (NFFO), ostensibly to support the nuclear industry, but the NFFO has recently been slanted towards renewables. Feed in tariffs have been used as incentives for the development of renewables in a number of European countries. The principles of feed in tariffs in some of these countries are outlined in Appendix III.

The renewable technologies include but are not limited to wind, hydropower, municipal and industrial waste, and biomass including the gasification of agricultural and forestry wastes. Many of these technologies are now generating electricity at about 6 US cents per kWh, which is comparable to conventional sources. Contracts under the NFFO have been financed by one of two ways:

1. Cash resources of a company have been used to finance the entire project (balance sheet financing).
2. Bank loans are secured mainly against future cash flows as well as against the physical assets of the project being financed.

Of course, there are many other possible financial mechanisms such as venture capital, joint ventures, public subscription, co-operative ownership and multilateral leveraged aid. There is a need to increase the availability of finance for renewable energy projects by improving the confidence of lenders in the technologies. The availability of finance will only be possible if lenders are aware of the importance of renewables. Lenders would need to know, for example, that there has been a significant growth in bioenergy over the last decade. Liquid biofuels are assisting in the reduction of emissions in the transport sector. Brazil has championed the use of bioethanol which now fuels more than 40 per cent of its rolling stock. Corn ethanol production in the United States has been ramped up to exceed that of Brazil's ethanol production from sugar cane.[3]

Lenders will also need to learn about the cost-effectiveness of carbon capture and storage. The geologic sequestration of CO_2 as part of the carbon capture and storage (CCS) process has proven potential to provide the needed reduction in atmospheric emissions. The capture of CO_2 from point sources would also create an important revenue stream as CO_2 could be sold for enhanced oil recovery. The benefit of CO_2 recovery to the economy would be significant, but it is important to examine how CO_2 recovery fits into the carbon balance and the long-term solution. (Koljonen et al., 2008) Techniques to assure that a sequestration site will retain stored CO_2 for geologically significant time periods include site characterization, modelling, and monitoring. Risks from failure of a sequestration site to perform properly include leakage to protected groundwater, impact on ecosystems, return to the atmosphere, and damage to other subsurface resources (gas reservoirs in particular). Assessment shows that maximum damage is limited and moderate, and that risks can be reduced by implementing a permitting process that favours high quality sites with adequate monitoring (Hovorka, 2009).

Cost-effectiveness of Energy Systems

A number of financial incentives should be employed to promote cost effective energy systems. Supply curves provide an approach to understanding the development and pace at which renewable energy systems will become more cost competitive and thus increase their value to energy systems. In general, it is difficult to compare data and findings from renewable energy supply curves, as there have been few studies using a comprehensive and consistent approach employing comparable methodologies. Furthermore, most renewable energy cost curve studies focus on single, or just a few, renewable energy resources. Some of these investigations have combined multiple technologies/resources applying a universal methodology (de Vries et al., 2007).

Let us examine one example. The supply curve for Germany for 2030 (Scholz, 2008) shows PV becoming available above US$200/MWh, whereas wind for example has a large potential with a lower cost of US$100/MWh. Nevertheless, once the cost level is reached where PV increases in market share, its potential improves and it becomes more competitive. This study also reinforces the significance of technological development in the case of PV, as the supply curve for 2050 shows that at that point of time costs are expected to go down at a scale that its full potential becomes available under around US$200/MWh. On the other hand, in the case of wind, the cost gap between 2030 and 2050 is considerably smaller and starts to widen only when approaching the maximum technical potential.

In the same context, in Europe as a whole the trend is similar in terms of the characteristics of supply curves with regard to the gap between 2030 and 2050 cost curves for these technologies. This means that it is not easy to predict the effect of new technological changes and development on some renewable energy technologies in the long term (Scholz 2008).

There have been several studies on the relationship between energy development and sustainability. Needs (2009 a, b) conducted quantifiable external costs for several electricity generating technologies. Estimation of external impacts and their valuation include considerable uncertainties and variability, as illustrated in Figure 18.1. Surprisingly, nuclear power received a high ranking in terms of its impact on human health. No considerations were given to the capability to handle nuclear terrorism in light of the proliferation of nuclear plants.

Most renewable energy systems should be cost effective in the long term, when the huge challenge of climate change is taken into consideration. Indeed, the monetary evaluation of the impacts of the changing climate is difficult. To a large extent the impacts manifest themselves slowly over a long period of time. In addition, because of the complexity of ecosystems, the impact may not only be felt at the geographic location of the polluter, but thousands of miles away.

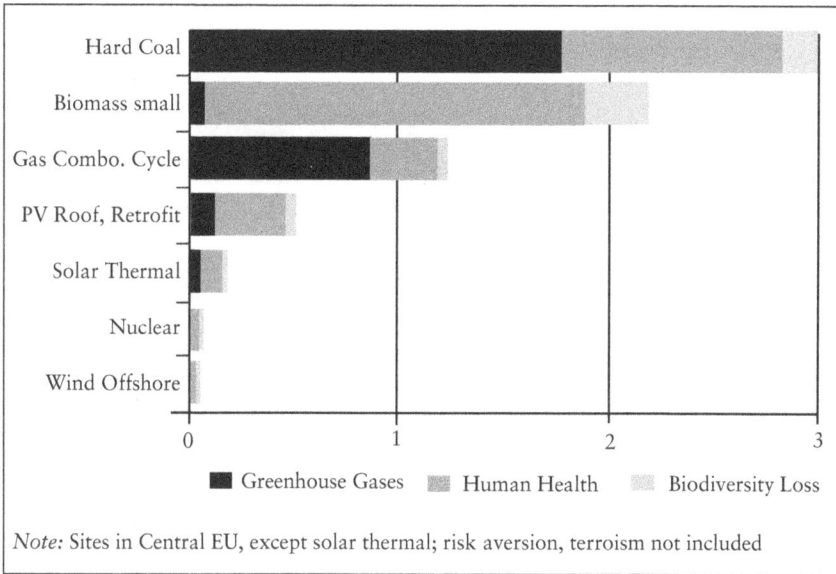

Figure 18.1 Comparison of External Costs of Various Energy Technologies
Source: Needs 2009 a, b

Climate Change and Mitigation: the role of the individual and the community

Thousands of miles away, sea levels may rise. New evidence indicates that sea levels worldwide could rise by nearly 100 cms by 2100. Presently, the coastal areas of a range of countries from Bangladesh to the Maldives, Liberia, Bahamas, to Alaska, USA are being threatened by rising sea levels. Further, since the end of the eighteenth century, the amount of carbon dioxide in the sea has increased by approximately 30 per cent, increasing the acidity of sea water which has a deleterious effect on some marine life, particularly shellfish and corals. These two factors, rising sea level and increasing acidification of the oceans, will have to be addressed with urgency if the world as we know it today is to be maintained for future generations.

Many developing countries, exemplified by Bangladesh, have begun to address the negative effects of climate change through the provision of clean energy such as solar. This strategy is also intended to improve the quality of life, as was accomplished with mobile telephones. This energy deployment pathway embodies interaction at the micro level (the community) the macro level (the nation) and the mega level on a global basis.

Recent erosion along Alaska's Arctic coast caused in part by rising sea level. Also to be noted is the collapsed block of ice-rich permafrost. Coastal erosion has increased significantly in Alaska along the Beaufort Sea in the last decade due to global warming

Ruminant animals, such as cattle are major producers of methane. Methane is a powerful green-house gas produced both naturally and through anthro-pogenic activities. The production of methane is a stimu-lant in effecting climate change

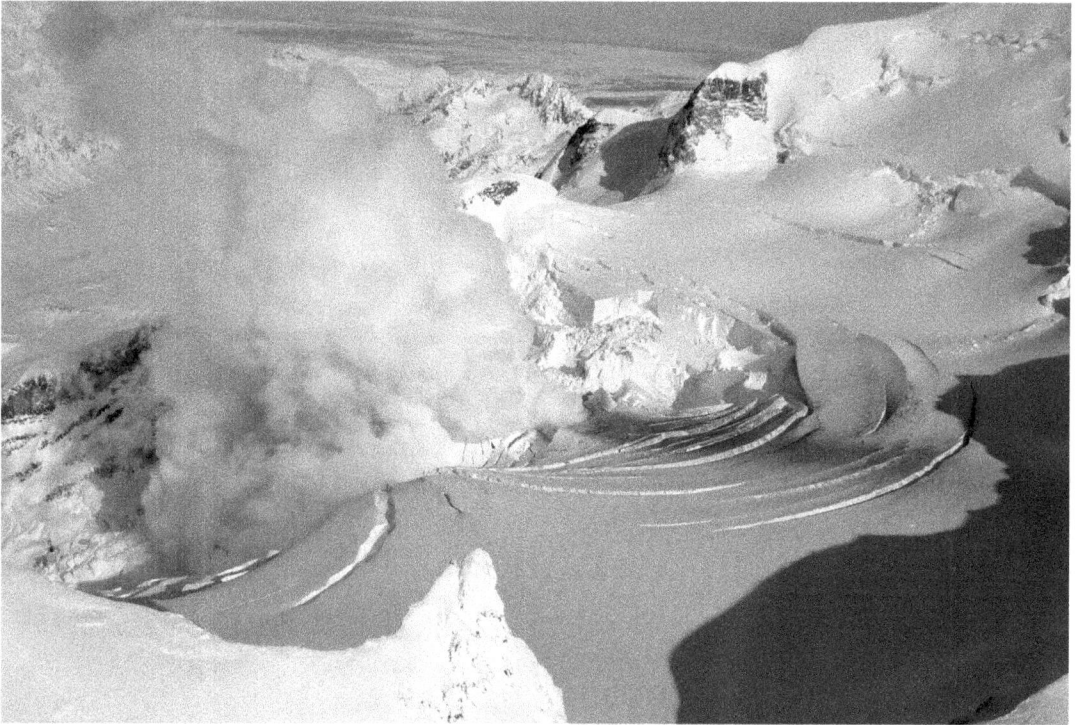

Retreating glaciers are a direct symptom of global warming

Frequent flooding in low line coastal areas such as parts of Myanmar is an effect caused by climate change

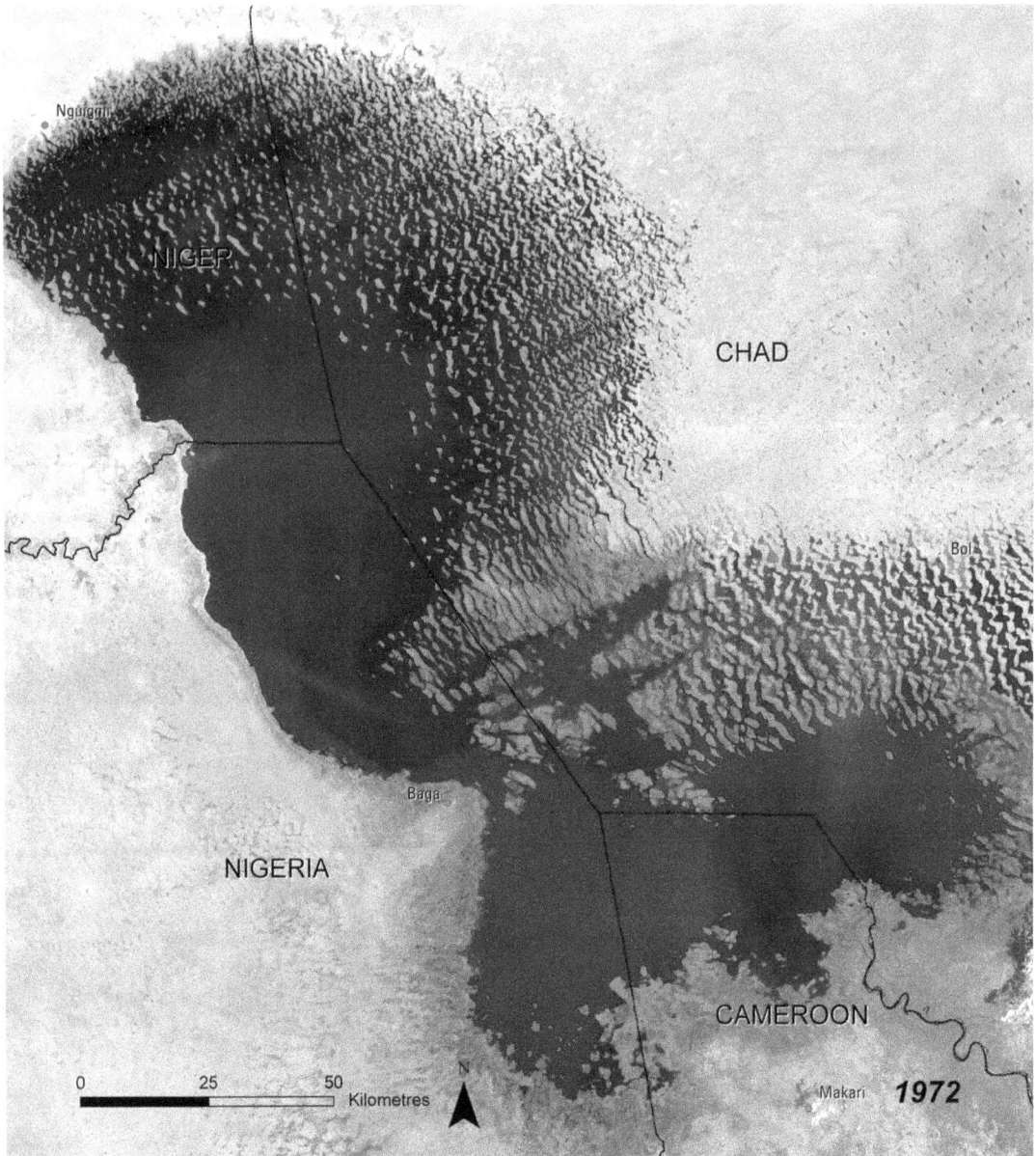

Lake Chad, 1972

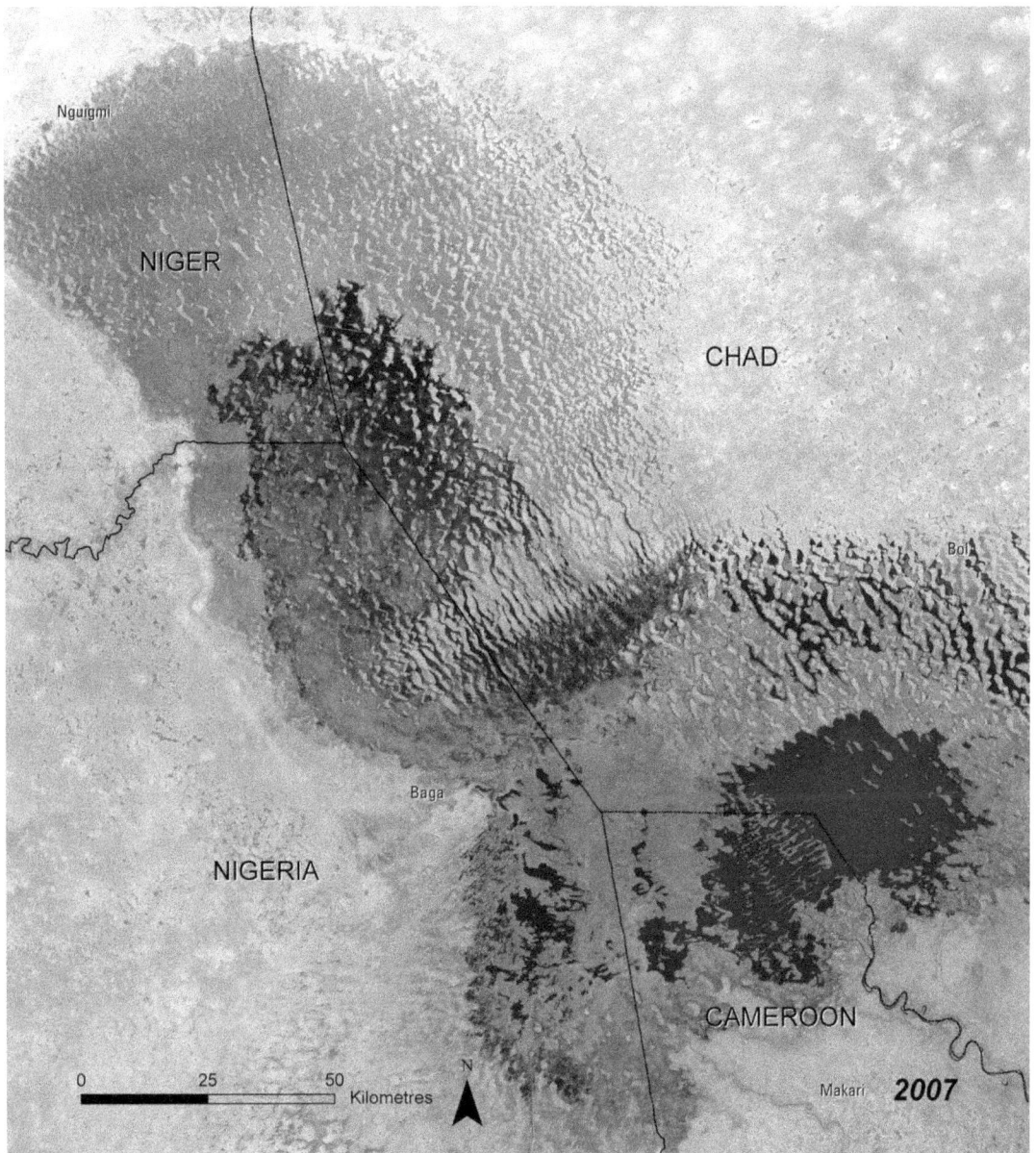

Lake Chad, 2007

Lake Chad, a freshwater lake located in central Africa at the junction of Chad, Cameroon, Nigeria and Niger. Lake Chad has shrunk dramatically over the last four decades due to a decrease in rainfall and an increase in the amount of water used for irrigation projects. Its surface area was 25,000 sq kms in the early 1960s and is now less than 1,100 sq kms

Warming of the ocean increases the ability of its waters to absorb carbon dioxide decreasing pH causing acidification which has a deleterious effect on the calcareous skeletal material of coral reefs

In many countries such as parts of Africa drought conditions have been caused by climate change

The process starts at the individual and community level, where social habits, anthropology, and micro economics become intimately intertwined and integrated with national policy and a global response. Thus the village becomes the starting point for an evolutionary process that provides a widespread and equitable distribution of energy, improved energy efficiency, and importantly, enhances the quality of life for all citizens. Distributed electricity in discrete areas and villages, as well as electricity generation in individual households will find solar, and to a lesser extent wind, to be appropriate technologies. Hence the rise in the use of renewables will be conditioned, in part, by individual and community action. Social factors and local planning will become cardinal inputs in the worldwide thrust towards a better energy future.

Carbon storage and sequestration technologies will receive much attention and developmental activity over the next two decades.[4] It will involve the use of reservoir capacity with pore space in areas where the geological strata, primarily marine shales, will act as seals. There will be legal issues as to who has ownership of the pore space in the sub-surface reservoirs. The technology will require global acceptance. The monitoring and verification of carbon storage and sequestration sites is crucial to public acceptability, as well as the requirements of industry investors and financial markets.

International Agreement for Mitigating Climate Change

The Copenhagen Summit of December 2009 recognized that billions of dollars will be required to mitigate the possible existential catastrophe that has been caused by climate change. Much was expected from the conference, but the agreement reached in Copenhagen proved only to be a beginning, not a breakthrough. The most tangible accord reached was the commitment made not to increase global temperatures by more than 2°C by the year 2100. The accord also recognized the critical impacts of climate change and the potential impacts of response measures on countries particularly vulnerable to the adverse effect of climate change.

With the evidence that there is a significant disparity in the ability of developed and developing nations to fund efforts in alleviating climate change, US$100 billion was committed by developed countries to finance the mitigation efforts of developing countries up to 2020. Funding for adaptation will be prioritized for the most vulnerable countries, such as the least developed developing states and Africa.

It is evident that gaining agreement and measures to mitigate climate change is a daunting task. Not only will it require large investments to control and curb emissions, but it will require cooperation from emerging developing nations such as China whose relative contribution to carbon emissions is growing rapidly with the progression of industrialization.

One may contend that a sustained effort towards a reduction in carbon emissions will not come only from International Treaties, the quantifiable results of which are not easily verifiable and hence not properly reportable. Indeed, positive impacts on climate change will be predicated not only by regulation but also, perhaps more importantly, by the widespread implementation of greener energy technologies. Therefore a strong case can be made for significant expenditure, on an international basis, to stimulate the urgent and wider use of cleaner fuels for transport and electricity generation.

A Sustainable Energy Future

There is a long list of new and potential energy sources, such as hydrogen, methanol, new biomass sources, fusion, safer nuclear fission and the large-scale development of fuel cells for electricity generation, as well as electric and hybrid cars for transportation. However, the traditional energy sources – coal, oil, natural gas and nuclear energy – will be needed for some time to come. The preferred option is not to replace existing energy sources by renewables on a gigajoule for gigajoule substitution, but rather to perform many energy services with much less use of primary energy.

Given the long lead times and the long life of capital energy assets, we are looking ahead in decades rather than years. Fifty years in the future is within the lifetime of our children. We should match this time horizon with the time scale of the political and bureaucratic process which emphasizes appropriate policies and financial incentives.

One of the most exciting advances for a sustainable energy future is carbon capture and storage. Before this process is accepted as a dominant part of the solution, it is important to consider the possibility that hazards resulting from this action will create unacceptable damages.

A sustainable energy future not only addresses alternative sources of energy and carbon control and storage, but also population control.

Indeed, further conceivable increases in energy efficiency are not only being offset by greater energy consumption per person, but also by the growth in population. Public education is a critical strategy in developing a sustainable agenda. Scientists, technocrats and administrators in developing countries that need to import energy should urgently communicate the necessity for population control and energy efficiency to the people.[5] If the communication is clear and simple, it is likely that many more persons will become interested in learning about the need for renewable energy options.

Simply put, the stabilization of polluting emissions at present levels cannot be achieved without the development of competitive non-carbon renewable energy sources. There also needs to be restraints on individual lifestyles that require high levels of energy. The harnessing and use of energy has become as important in our lives as air, earth, and water. Through technological improvements and breakthroughs, we will need to make energy sustainable in order to create a more secure future for ourselves and our children.

References

de Vries, B.J.M., van Vuuren, D.P., and Hoogwijk, M.M. (2007) "Renewable Energy Sources: Their Global Potential for the First-Half of the 21st century at a Global Level: An Integrated Approach". *Energy Policy* 35 pp. 2590–2610

Hovorka, S. (2009) "Risks and Benefits of Geologic Sequestration of Carbon Dioxide – How Do the Pieces Fit?" Abstract, AAPG Convention, Denver.

Koljonen, T., Flyktman, M., Lehtilä, A., Pahkala, K., Peltola, E., and Savolainen, I. (2008) "The Role of CCS and Renewables in Tackling Climate Change" GHGT-9 November 16–20, 2008 Washington D.C. Published in *Energy Procedia*.

Kump, L., Ridgwell, A., and Panchuk, K. (2009) *The Carbon Cycle and Fossil Fuels* Abstract, AAPG convention, Denver.

Lindley, D. (1996) Financing the UK's Renewable Energy Boom *International Journal of Global Energy Issues* 8 (5/6) pp. 425–435.

Needs (2009a) Policy use of NEEDS results. Project: New Energy Externalities Development for Sustainability. Retrieved from: http://www.needs-project.org/docs/Needs.pdf

Needs (2009b) A Summary Account of the Final Debate. Project: New Energy Externalities Development for Sustainability. Retrieved from: http://www.needs-project.org/docs/Annexstampa.pdf (read in November 2009).

Phillips, M. (1991) The Least-Cost Energy Path for Developing Countries: Energy Efficient Investments for the Multilateral Development Banks. Washington DC: International Institute for Energy Conservation.

Rathmann, R., Szklo, A., and Schaeffer, R. (2010). "Land Use Competition for Production of Food and Liquid Biofuels: An analysis of the Arguments in the Current Debate". *Renewable Energy* 35(1) pp. 14–22.

Soimakallio, S., Antikainen, R., Thun, R. (Eds.) (2009a) Assessing the Sustainability of Liquid Biofuels from Evolving Technologies. Technical Research Centre of Finland. VTT Research Notes 2482. 220p.+app.41p.

Scholz, I. (2008) Erneuerbare Energien für Elektromobilitat: Potenziale und Kosten (Renewable Energy for Electromobility: Potentials and Costs). Presentation. Deutches Zentrum für Luft und Raumfahrt.

Epilogue

> Of all energy sources, oil has loomed the largest and most problematic: because
> of its central role, its strategic character, its geographic distribution, its recurrent
> pattern of crisis in its supply . . . Creativity, dedication, entrepreneurship, inge-
> nuity, and technical innovation have coexisted with avarice, corruption, blind
> political ambition, and brute force . . . Ours truly remains the age of oil.
> – Daniel Yergin

JAMAICA HAS BEEN HIGHLIGHTED IN the book as an example of a
small island developing state which imports its energy. In general, energy
is well distributed in Jamaica. More than 90 per cent of households are
within 150 metres of the electricity grid and have access to electricity
service. The value added by production in all Jamaican industries continues to be energy intensive with an energy input of approximately 10
US cents for every US$1.00 equivalent of gross domestic product (GDP).
Currency devaluation as well as increased consumption has significantly
increased the cost of energy imports to Jamaica.

On an end-use basis, approximately 92 per cent of commercial energy
in Jamaica comes from petroleum, the remainder from bagasse,
hydropower and coal. The bauxite/alumina sector uses approximately
one half of the energy provided in Jamaica, the transport sector approximately 24 per cent and the electricity sector approximately 26 per cent.
The bauxite/alumina industry, coupled with tourism, as well as the social
and cultural habits of the people, results in Jamaica being highly energy
intensive. Yet the country has not yet discovered any domestic source of
oil and gas, or coal, and is heavily dependent on these imported fuels.
For this reason, among others, Jamaica needs to diversify its energy base
and utilize indigenous resources where possible.

In the context of globalization, after the turn of the century there will very likely be greater diversification, more interfuel competition, and an energy industry more involved in all aspects of energy, rather than only the traditional resources of oil, gas and coal. While these natural energy resources are required for development, it is the human resources of a country that will convert them into genuine economic value. Human resources, coupled with a strong energy base, will be an important factor in the dynamic move to greater national energy independence.

Four factors are important in the future development of energy worldwide, which will of course affect Jamaica:

Availability of resources: It now appears that although oil, gas and coal are finite resources their supply will be more than adequate for at least another century. The major issues in energy should be the security of supply and not necessarily the exhaustion of resources.

Availability of technology: This will produce new systems of generation, introduce new efficiencies and enable oil, gas, coal and, ultimately, nuclear to provide cleaner and more efficient electricity. For example, the price of oil has not increased in real terms adjusted for inflation, yet its supplies increase unabated. Technology drives supply by reducing exploration cost and risk, increasing recoveries, and improving consumption efficiencies. Technology and operating efficiencies, more so than commodity price dynamics, will extend the reserve life of all energy resources.

Environmental protection: In the foreseeable future, carbon dioxide emissions will become part of the political agenda, both nationally and internationally. Environmental issues make renewables an imminently more attractive source of energy for the future.

Market dynamics: Competition among fuels, a growing energy resource base and the introduction of many new independent producers in electricity generation are creating new marketing dynamics that affect the commercial strength of different energy sources as well as extending their reserve life.

In 1990, Jamaica began the deregulation of its energy industry in tandem with liberalization, removal of trade barriers, and the reduction of distortions in prices. This strategy has among its goals:

- diversification of energy resources
- improving energy efficiency and conservation measures
- increasing private sector participation in energy production

The involvement of the private sector in energy has had an important impact on power generation as well as on the marketing of petroleum products. A number of independent power producers now provide energy to the utility company.[1] The Jamaican public utility company is itself mostly privatized (80 per cent private owned and 20 per cent government-owned). The oil refinery also acquired a Joint Venture Partner (PDVSA), making the private sector the major participant in an important energy service industry. Privatization is beneficial to the nation as long as companies do not require government guarantees and provide least-cost alternatives, use reliable technology, are competitive and economically viable. The entrepreneurial mode requires adequate regulation and in response to this need the Office of Utility Regulation (OUR) was established in 1996.

The new initiatives are supported by an energy policy that has among its requirements:

- ensuring adequate and stable energy supplies at the least economic cost in a deregulated and liberalized economy
- encouraging efficiency in energy production in order to reduce the energy intensity of the economy
- diversifying the energy base, encouraging the development of indigenous energy resources, and, at the same time, ensuring security of supply
- reducing adverse environmental effects caused by the use of energy, including the deleterious effects of the unmanaged use of fuelwood
- recognizing that energy is a critical factor in industrial growth; the energy sector must be complementary to the industrial sector

The Energy Policy is bolstered by an Energy Plan which is expected to make Jamaica attain a performance similar to that of a developed country by 2030. See Appendix I for further information on Jamaica's Energy Policy, and Appendix II for the fundamentals of the Energy Plan. The reader will note that the Energy Plan addresses environmental sustainability.

Appropriate choices regarding environmental sustainability and pollution can be made as we gather more precise quantification of the likely impacts of climate change mitigation. There are at least two viable options to lessen pollution. The first is to ration the consumption of oil and other fossil fuels in the ratio of their contribution to pollution from carbon dioxide. The second is to charge the producer of carbon dioxide (and other pollutants) with the social value of the emission, and using the 'tax' to address environmental problems.

Defining the social cost of a pollutant is difficult but a body of economic researchers is now addressing the value of social benefits or costs from environmental pollution. Soon, new renewable energy technologies will be encouraged by their lesser external costs. The start of this transition to more renewable energy use has among its components technology development, project demonstration, research and development support, tax credits, technology transfer, training and public education. As we adapt to the changing requirements of the energy sector, we should seek and identify creative energy solutions that will help to bolster economic growth (Dasgupta, 1995).

Many of us have associated both economic growth and environmental degradation primarily with oil, but energy companies should realize that oil will not experience the greatest growth among fuels. Over the next fifteen years one can predict increases in fuel use of the following order:

Oil	–	15 per cent
Coal	–	25 per cent
Natural gas	–	100 per cent
Renewables	–	300 per cent

Thus the energy company of the future will be weighted towards hydropower and new renewables. Also, energy companies will be more integrated, with less specialization in one type of energy. The energy industry will see:

- a trend towards greater privatization, deregulation and fiercer competition in the marketplace
- a shift towards renewables and clean fuel technologies
- more power generation being done by independent power producers.
- the honing of entrepreneurship, with emphasis on performance-based management, reduction in operational costs, and an unbundling of goods and services through specialization in the marketplace
- significant increases in the use of sophisticated informational technology, making work easier and faster

The energy industry is still immature. Over the next 30 years efficiency, diversification, globalization and technological change will increase in impact and importance. The industry will renew itself, mature, and with a new agenda will hopefully prove that its best days are yet to dawn.

References

Dasgupta, P. (1995) "Poverty, Population, and the Environment". In S. Hanna and M. Munasinghe (Eds.) *Property Rights and the Environment: Social and Ecological Issues*. Washington, DC: Beijer International Institute of Ecological Economics and The World Bank.

Yergin, D. (1991) *The Prize*. London: Simon and Schuster p. 788.

Policy as an Important Tool in Energy Management

The Case of Jamaica

Policy Issues and Recommendations: 2010–2030

Energy Supply and Security

Policy Issues:

- Security of Energy Supply
- Diversification of Energy Types
- Need to pursue Oil and Gas Exploration

Petroleum Industry

Policy Issues:

- Maintain a Competitive Petroleum Industry with Industrial Harmony
- Need to ensure appropriate behaviour in a Competitive and Deregulated Environment
- Need for Intervention in Time of Disasters
- Need to prevent Mergers and Acquisitions that could Impair Competition
- Petroleum Refining Capacity
- Need for a relevant Petroleum Reference Price

PetroleumTax

Policy Issues:

- There is need to review the Petroleum Tax Regime
- Need for Sustainable Source of Funds to Support Road Maintenance

Electricity Sector

Policy Issues:

- Need for a Transport and well-regulated Electricity Market
- There are concerns regarding responsibility for the development of least cost expansion plan (LCEP) for electricity within the context of a privatized electricity sector and liberalized generation market
- Need for Heat Rate to be Consistent with International Standards
- Need for Prescribed Protocols for the supply of Electricity to the National Grid
- Need to Optimize Efficiency in Transmission and Distribution of Electricity
- Need to Ensure Accuracy of Meters and Billing Systems

Rural Electrification Programme (REP)

Policy Issues:

- There is a need to provide Electricity to remote Communities and Marginalized Groups
- Transfer of Assets Owned by the REP
- Need for more Competition in the Electricity Market

Transport Sector

Policy Issues:

- Need for Small Engine Size vehicles
- Need for more diesel-powered Engines versus Gasoline
- Octane Enhancement of Fuel
- Introduction of Biofuels
- Need for increased use of Flexi and Hybrid Vehicles
- Need for Greater use of Public Transportation
- Development of Renewable Energy Resources
- Need to increase use of Renewable Energy to Complement Fossil Fuel
- Need for Institutional Focus for Development of Renewable Energy
- Need to expand use of Solar and other forms of Renewable Energy at the Household Level
- Energy Conservation and Efficiency
- Need to Conserve on, and Improve Efficiency of, energy use in the Domestic Energy Sector

Energy Fund

Policy Issues:

- Need for a Dedicated Energy Fund to Finance Energy Conservation, Efficiency, Renewable and related Projects

Institutional Arrangement

Policy Issues:

- There is need to rationalize the Institutional Arrangements for Policy Development and Implementation, Regulation and Monitoring Functions for the Energy Sector
- Need for Timely and reliable Information and appropriate Analytical Tools to Support Policy

Development and Planning

Policy Issues:

- Financing the Energy Investment

Appendix II

Fundamentals of the National Energy Plan
VISION 2030

THE ENERGY SECTOR REPRESENTS A critical component of any country in its impact on national development. Energy is an essential input into all production processes, and is fundamental to the provision of social services that contribute to the well-being of urban and rural populations. The modes by which energy is produced, distributed and consumed also have wide-ranging implications for the long-term sustainability of the environment. The energy sector plan also will have implications for other areas of national development including transport, tourism, urban and regional planning, agriculture and mining.

The structure of the energy sector clearly indicates the areas that present the main challenges for its long-term development and transformation. As indicated above, the bulk of Jamaica's energy resources are consumed in three (3) areas:

1. Transport sector
2. Bauxite and alumina
3. Electricity generation

The long-term planning for the energy sector therefore must be focused on these main areas, in order to achieve meaningful improvements. Reduction in the cost of electricity and other energy supplies must be a clear priority in the medium- and long-term.

Additionally, while the development of alternative energy sources including renewable energy sources will be an important aspect of the long-term diversification of the sector, it is likely that fossil fuels will remain the main source of energy for Jamaica over the planning horizon through 2030.

The overall goal for Jamaica is to develop an energy sector that can contribute to long-term economic competitiveness, improved quality of life and sustainable environmental management. This will involve addressing a wide range of separate challenges, including the following:

1. Increase in Energy Supply and Security:

Jamaica must plan to increase its supply of energy to meet projected increases in long-term demand. At the same time Jamaica must seek to increase its energy security to reduce its vulnerability to potential disruptions in energy supplies. Over the long term geo-political factors have posed significant risks to the continuity and cost of global energy supplies. The measures that can contribute to achieving long-term increase in energy supply and security include:

- Replacement of existing plant with more efficient generators
- Construction of new energy-efficient generating facilities on a phased basis to meet increased demand
- Reduction in dependence on imported petroleum through diversification of energy sources such as natural gas, coal and renewable energy sources
- Maintenance and enhancement of bilateral agreements with regional energy partners including Mexico, Venezuela and Trinidad
- Exploration for exploitable fossil fuel energy resources in Jamaica

2. Reduction in Energy Costs:

It is imperative for Jamaica's economic and social development for the cost of energy to be reduced significantly over the medium term. It is important to recognize that high energy costs are primarily associated with the cost of electricity generation, as Jamaica has relatively low gasoline prices and tax rates for a non oil-producing country. The reduction of energy costs will involve a number of measures including:

- Reduction in the cost of electricity generation in the public system and the bauxite and alumina industry through increasing the energy efficiency of the generating plant, switching to lower-cost fuels, or a combination of both
- Reduction of system losses in electricity transmission and distribution to international benchmark levels

3. Increase in Energy Efficiency in Supply and Demand:

Jamaica also must achieve significant increases in efficiency in producing and using energy, which will involve a range of measures including:

- Upgrading of the petroleum refinery to increase capacity utilization and output of lighter and higher-value refined petroleum products in order to replace imports and compensate for the potential switch from oil-fired to natural gas power plants
- Encouragement of more fuel-efficient vehicles in the transport sector including the use of diesel and bio-fuels
- Implementation of demand side management programme including the use of energy-efficient appliances, equipment, and building designs, setting and enforcing standards for public sector organizations, and public awareness and educational programmes
- Use of the Energy Efficiency Fund to support energy conservation

4. Social Equity and Environmental Sustainability:

The long-term development of the energy sector also must address the following social equity and environmental sustainability issues:

- Completion of rural electrification including deep-rural households
- Reduction in adverse environmental effects from use of fossil fuels in manufacturing, bauxite and alumina plants, power generation and other industries
- Use of cleaner production processes
- Reduction in vehicle emissions through a combination of regulatory mechanisms including the National Vehicle Emissions Standards
- Introduction of biodiesel and ethanol blends to replace methyl tertiary-butyl ether (MTBE) as fuel additive
- Compliance with international conventions on climate change and global warming
- Increase in contribution of renewable energy sources in Jamaica

5. Appropriate Policy and Regulatory Framework:

The energy sector will require an appropriate policy and regulatory framework to meet the range of challenges identified above, including the following:

- Regimes for pricing of electricity and petroleum products that will balance requirements for competitiveness with the long-term viability of the sector

- Appropriate tax and pricing structure for road users that reflect environmental costs and other externalities
- Institutional framework to coordinate policy with energy initiatives and provide integrated monitoring and enforcement of regulations
- Promotion of a market based approach and increased competition in the sector including a transparent procurement process for new capacity and sourcing from private producers
- Rationalization of the number of existing Acts governing the sector through the introduction of new modern industry legislation.

Appendix III

Fiscal Incentives

The Case of Feed-in Tariffs

FEED-IN TARIFFS MAY BE established to allow Independent Power Producers (IPPs) to provide energy to utilities at a set rate. This will allow IPPs to develop projects with a knowledge of the tariff to be provided by the utility company. In this regard, the regulating agency plays a critical role.

Experiences with Feed-in Tariffs in European Countries

Germany

Feed-in tariffs in Germany have been officially introduced since January 1, 1991 when the so called 'Electricity Feed Law' (EFL) came into force. In April 1998, the EFL was amended at certain points, while two years later – in April 2000 – it was decisively revised and replaced by a new Act called the 'Renewable Energies Law' (REL). The EFL of 1991 regulated the purchase and price of electricity generated in the territory of the Federal Republic of Germany from specified renewable sources (for example, hydropower, wind energy, solar energy, landfill gas, sewage gas, and biomass). Excluded from the law were:

(i) installations using resources other than wind or solar energy with an installed capacity of more than 5 MW, and

(ii) installations in which the Federal Republic of Germany, a federal state, a public electricity utility or one of its subsidiaries held shares of more than 25 per cent.

238

The EFL obliged the grid companies to purchase renewable electricity from eligible sources and to pay the producers concerned an annually fixed feed-in tariff. For power generated from solar or wind energy, the tariff was set at 90 per cent of the average electricity utility rate per kWh of all final consumers charged over the last but one calendar year. Hence, if consumers had paid, on average, 10 ct/kWh in 1993, a farmer exploiting a wind turbine received 9 Eurocents for every kWh fed into the grid in 1995. For electricity produced from other eligible sources of renewable energy, the corresponding feed-in tariffs were set at lower rates – i.e. either 80 or 65 per cent of the average consumer price – depending on the output capacity of these sources.

As part of the EFL amendment of 1998, a so-called 'hardship clause' or 'cap' of 5 per cent on the purchase obligation of grid companies was introduced. This clause stipulated that if the amount of renewable electricity to be supported by a power utility surpassed 5 per cent of its total deliveries in one calendar year, the upstream system operator had to reimburse the costs of purchasing additional renewable electricity until it also reached the 5 per cent ceiling in its grid area.

Actually, this implied that beyond the 5 per cent threshold, the utility was no longer obliged to purchase renewable electricity offered to the grid. Besides other support mechanisms such as preferential planning guidelines and lower interest rates for loans to invest in wind turbines, the EFL was particularly successful in stimulating the installed capacity of wind energy in Germany. After the law was passed, this capacity more than doubled each year during the period 1990–95. The average annual growth of installed wind power slowed down to some 40 per cent in the second half of the 1990s, but in absolute terms the installed capacity expanded substantially from 1,100 MW in 1995 to 6,100 MW in 2000, making Germany the world leading country in wind energy.

Denmark

Between the mid-1980s and the late 1990s, feed-in tariffs in Denmark have played a prominent role in stimulating electricity generated from renewable energy sources. Due to the introduction of the new Electricity Supply Act on January 1, 2000, however, this role will be less important in the coming years as – starting from 2003 – feed-in tariffs will apply only for existing renewable energy installations during a temporary, transition period of 10 years while for new installations a more market-oriented mechanism – i.e., tradable green certificates – will be used as the main instrument to encourage renewable power production.

This changing role of feed-in tariffs in Denmark before and after the year 2000 is discussed below.

Before 2000, the system of feed-in payments in Denmark differed by type of renewable energy technology. For electricity from biomass fed into the grid, the utilities paid a feed-in tariff based on the principle of avoided costs, depending on the specific time at which the electricity was supplied. As a result, feed-in tariffs for biomass varied between 2 and 13 ct/kWh in the late 1990s, with the average price received by independent power producers amounting to approximately 4.3 ct/kWh. For electricity from wind power, on the other hand, the feed-in tariff was based on 85 per cent of the consumer price of electricity in the given distribution area. Hence, as the tariff depended on the location of the wind turbine, it varied between 3 and 5 ct/kWh in the late 1990s.

Denmark has opted for a system of promoting renewable electricity in which the role of feed-in payments is gradually reduced in favour of more market-oriented instruments such as competitive pricing and tradable green certificates.

Spain

Since the mid-1990s, feed-in tariffs have been a major instrument to promote renewable electricity in several other countries in West Europe such as Austria, Greece, Italy, Luxembourg, Portugal and – particularly – Spain. In Spain feed-in tariffs were introduced in 1994 by means of the so-called 'Royal Decree 2366'. In addition to the market price of electricity, producers of renewable energy in Spain receive a premium feed-in tariff, which amounted to some 3 Eurocents per kWh in 2000 for most renewable energy sources and even 36 ct/kWh for small-scale solar plants. Since the mid-1990s, Spain – together with Germany and Denmark – has belonged to the group of countries with the highest feed-in payments to renewable power producers.

Evaluation

Based on the experiences of the European countries outlined above – supplemented by some general, theoretical reflections from the literature – the performance of feed-in tariffs in promoting renewable electricity is evaluated with regard to the following criteria:

- Investment certainty
- Effectiveness
- Efficiency
- Market compatibility and competition
- Administrative demands

Investment certainty: The core argument to apply feed-in tariffs is that they offer a high level of investment certainty to independent (risk-averse) producers of renewable electricity by guaranteeing a fixed price for each kWh of power fed into the grid over a certain period, for instance 5–15 years. As noted above, however, this certainty may particularly apply to the short or medium term as in the long run fixed feed-in tariffs may be unsustainable either because of high cost inefficiencies involved or because they are not compatible with a liberalized, competitive market and a system of harmonized, renewable energy policies within the EU.

Effectiveness: Owing to the investment certainty offered by fixed feed-in tariffs, the latter have been very effective in promoting renewable electricity, notably wind power in countries such as Germany, Denmark or Spain. Their effectiveness, however, depends largely on the particular level of the tariffs set as well as on other factors such as the production costs involved, the existence of other promotion schemes, administrative procedures, natural conditions or other specific characteristics at the local, regional or national level. Moreover, it should be stated that, apart from wind power, feed-in tariffs seem to have been far less effective in encouraging other forms of renewable electricity.

Efficiency: The major criticism with regard to feed-in tariffs is that they have failed to be efficient in both static terms – i.e., able to ensure that electricity is generated and sold at minimum costs – and dynamic terms, i.e., able to foster innovations and, hence, further drive down costs. Compared to other promotion schemes, feed-in tariffs have generally failed to result in price reductions for renewable electricity. For instance, between 1994 and 1998, the minimum bid prices for contracts under Britain's auction model of support – the so-called Non-Fossil Fuel Obligation (NFFO) – fell 40 per cent in real terms, whereas feed-in tariffs in Germany and Spain remained more or less stable. This lack of cost/price efficiency of feed-in tariffs can be subscribed to the following factors.

Feed-in tariffs are generally fixed by a regulatory authority. This authority, however, usually lacks adequate, up-to-date information regarding the production costs of renewable electricity from a variety of different sources and technologies, notably in dynamic terms over a certain period.

Therefore, it is very hard to fix the 'right' price and to differentiate it adequately over time or by different types of renewable energy sources and technologies. Moreover, it may be unpopular and, hence, politically difficult to reduce feed-in tariffs as existing producers have strong economic interests in ensuring continued high feed-in payments.

As a system of feed-in payments is not based on direct competition – either among renewable power producers or between these producers and nonrenewable electricity generators – the incentive for innovations will, by definition, be less pronounced than under schemes based on competition.

Market compatibility and competition: Another major criticism of a system of feed-in tariffs is that it implies a distortion of free competition and, hence, that it is not compatible with a single, liberalized electricity market in Europe. Basically, there may be three problems with regard to this issue of competitive pricing and marketing.

Firstly, feed-in tariffs do not go together with competitive pricing between generators of green electricity, which may result in less efficiency in renewable power production. Secondly, a national system of feed-in tariffs that is eligible to domestic generators of green electricity only and excludes imports of renewable electricity may conflict with EU rules regarding non-discrimination of domestic versus foreign producers and free international trade among Member States. On the other hand, non-discrimination of producers and free international trade may lead to major imports of green electricity and major outflows of financial resources, which may be unacceptable for a country offering relatively high feed-in tariffs.

Administrative demands: A major advantage of a system of feed-in tariffs is that its administrative demands are, in principle, low and simple. However, as already indicated above, both the administrative demands and the informational needs of a feed-in tariff system will rise rapidly if a compensation mechanism covering all grid utilities is introduced:

(i) the system is extended from the national to the international level, and

(ii) the system becomes more fine-tuned and complicated in order to meet the efficiency conditions discussed above.

Overall, it may be concluded that a system of premium feed-in tariffs has shown to be an effective instrument to promote the generation of renewable electricity, notably to ensure a low-level market take-off of wind power at the national level.

In the longer term, however, such a system may become hard to sustain as it may suffer from some major drawbacks, especially when the generation of green electricity accounts for a significant share in total power production. These disadvantages refer particularly to the fact that a system of fixed premium prices tends to be costly, inefficient, distortive of competitive pricing and, hence, incompatible with the creation of a single, liberalized electricity market in Europe.

In the long run, the best way to encourage renewable electricity within a free European market is probably either to internalize the external costs and disadvantages of non-renewable energy sources – e.g., by means of taxation – or to introduce market-conform instruments such as a well-functioning system of tradable green certificates (where the price of these certificates accounts for the social and environmental benefits of renewables compared to nonrenewables).

However, it may take quite some time – if ever – before either one of these 'best means' (or a combination of both) will be achieved. In the meantime, feed-in tariffs can and will be justified in several European countries as the best alternative instrument to encourage the generation of a certain amount of green electricity, notably when this amount is still small.

Appendix IV

Rational Energy Use

The Wisdom of Efficiency and Conservation

BY PROMOTING ENERGY EFFICIENCY WE can reduce the rate of growth of energy demand and avoid the need to make new investments in electricity generating plants. Among the approaches which the utility company will consider in the future is providing the consumer with an incentive to install high-efficiency lighting and appliances.

One of the most important procedures is to circulate success stories of energy efficiencies among Jamaicans, to expand the experience gained in conservation, and to duplicate them in places where it is appropriate. At the same time, we need projects which can be easily executed using the expertise which exists in Jamaica. This will stimulate technical persons and business entrepreneurs to implement energy efficiency and renewable energy projects whilst providing additional employment.

High consumption in the domestic sector takes place in areas of lighting, refrigeration, space cooling, water heating, cooking and entertainment. Among the popular appliances were stoves, microwave ovens, toasters, percolators, water heaters, fans, refrigerators, air conditioners, and lighting. Consumers were willing to invest in energy conservation but they wanted to know the savings and the price incentives. This section addresses everyday energy use and suggests simple steps to effect energy savings in the home.

Kitchen

Some activities in our lives are optional, but not cooking and eating. Thus what happens in the kitchen is absolutely necessary, and occurs every day. For that reason, a careful examination of how energy is used in the kitchen is vital to an overall economy plan. Also, because the kitchen is an everyday venue for all members of the household and is the place where much of the shared life takes place, the development of new habits in the kitchen may raise the family consciousness about energy use in general, and so carry over into other aspects of energy saving.

Opening the refrigerator door is costly. Cold air rushes out as soon as the door is opened. The more frequently the door is opened, the more cold air escapes. When you stand with the refrigerator door open, thinking about what you would like to have, you are running up the cost of that snack. Do your best to imagine what is inside before you open the door, and then go directly to it. Try to teach your children this habit too.

After coming home from the supermarket, empty all the shopping bags on the counter, put all the items that need refrigeration in one place, then open the refrigerator door.

The frost-free feature is a convenience, but a standard refrigerator that must be defrosted by hand, say once a month, will use less electricity. If you want the frost-free convenience, look for a model with a power-saver switch. It can cut operating costs by as much as 15 per cent. When you are refrigerator shopping, look for the yellow Energy Guide which gives the manufacturer's information on average annual operating costs.

If you have a separate freezer, you should realize that it is the most expensive electrical appliance to operate. A manual defrost 14 cubic foot model will use about 100 kilowatt hours of electricity per month. You will use 40 per cent more electricity with an automatic defrost model. Weigh that against the modest effort of defrosting. Try to keep your freezer as full as possible. Most foods will retain the cold better than empty air, giving a more economical operation. A last thought on freezers – upright models are more convenient to use but every time you open the door the cold 'falls out'. Chest-type freezers are much more frugal in operation.

Hot Water

Hot water is important in three areas of the home: kitchen, bathroom, and laundry. These combine to make the water heater one of the major energy consumers in the home, whether your fuel is gas or electricity. In the kitchen sink you should wash as many utensils as possible at one time. In the bathroom a quick shower takes about half as much water as a tub bath. Regard soaking in the tub as an occasional luxury, and the quick shower as a frequent sanitary necessity.

Be sure to turn off the faucet all the way when you have finished using the sink or tub. If it still drips you are losing money. One drop per second from a hot water faucet is 800 litres a month, 9,600 gallons per year.

Another option is to install a solar hot water system. These cost between J$80,000 and J$110,000 (in 2009 dollars) to install and will pay for themselves in 4–5 years, when compared with electric water heaters which cost less but use electricity throughout their life cycle.

Life Cycle Costing

When the time comes to make a new purchase, perhaps to replace an ageing appliance, try to look beyond the purchase price by considering how much it will cost to use the appliance. Comparing the life-cycle cost, which is what it costs over the appliance's entire lifetime, is an easy and effective way to calculate which appliance is the most energy efficient.

This is how it is done. Estimate how long the appliance will last. Estimate the life-long operating cost of the appliance. This includes both energy and maintenance costs. Finally, add the purchase price and the lifetime operating costs together to obtain the total life-cycle cost for the appliance.

Here is an example. Refrigerator 1 costs $30,000 and $5,000 per year to operate for a lifetime estimated to be 15 years. Therefore, its life-cycle cost will be $105,000. Refrigerator 2 costs $35,000 to purchase and $4,000 per year to operate for 15 years. Its life-cycle cost is $95,000. Hence, although Refrigerator 2 has a higher purchase price it is a better investment overall.

Lighting

An effective approach is to reduce the wattage of light bulbs. This can be done in three ways:

- by using lower-wattage incandescent bulbs
- by replacing a number of lower-wattage bulbs with one higher-wattage bulb
- by replacing high-use incandescent bulbs with fluorescent and compact fluorescent bulbs

Lower-wattage bulbs should be used in halls, vestibules and other places where no close-up work or reading occur. In areas that need better lighting, it is often more efficient to use a higher-wattage bulb than a number of lower-wattage bulbs. For instance, one needs five 25-watt bulbs to get the same amount of light available from one 100 watt bulb, and the five 25-watt bulbs use 25 per cent more electricity.

Fluorescent lights use nearly one-quarter the amount of electricity as their incandescent counterparts. For example, a 15-watt fluorescent bulb produces as much light as a 60-watt incandescent bulb. Although the initial cost is greater, each fluorescent bulb will last about 7.5 times longer than incandescent bulbs (Figure AIV.1).

Also remember that long-life incandescent bulbs are more expensive and less efficient than standard bulbs of the same wattage.

At the same time, efforts should be made to use electricity for lighting. Incandescent and fluorescent light bulbs are ten to twenty times more efficient than kerosene lamps which are still a common source of light in rural Jamaica.

Lamp A	Lamp B
Incandescent Light Fixture	Fluorescent Light Fixture
Power Requirements: 25 watt Illumination: 385 lumens Total fixture cost (US$): 5.00 Replacement bulb cost (US$): 1.00 Average bulb life: 1,000 hours	Power Requirements: 8 watts Illumination: 400 lumens Total fixture cost (US$): 30.00 Replacement bulb cost (US$): 2.50 Average bulb life: 7,500 hours

Figure AIV.1 Comparison of Incandescent versus Fluorescent Bulbs in terms of Cost and Efficiency

The focus in energy saving should be on lighting and the refrigerator-freezer. As much as 25 per cent of the electrical costs in the average household is accounted for by the refrigerator and another 25 per cent by lighting.

Lastly, do not forget the simplest solar energy device of all, the solar clothes dryer. It is excellent and efficient, this backyard clothes-line, and it saves significantly when compared to an electric clothes dryer.

Comfort and increased standards of living can be had through investments in energy conservation. It should be approached positively and sensibly.

Energy efficiency is required in every aspect of energy use including households, commercial activities, hotels and the hospitality industry, manufacturing, and the transportation sector.

In households, the first step is to use compact fluorescent lamps for lighting and as a replacement for all incandescent units. This replacement would save the need for somewhere between 30 and 40 megawatts of electricity if it was done in totality. Reasonable market penetration will result in a saving of about 22 MW. Other household efficiencies include purchasing the most energy efficient equipment such as refrigerators, freezers, microwave ovens and air-conditioning units. Attention should be paid to the use of electric irons as infrequently as can be accommodated and a reduction in the electric current used by "vampires" such as cable boxes, mobile phone chargers and television sets which should be plugged out when not in use. Where hot water is a requirement, solar water heaters should be the modus operandi. In order to effect energy efficiency in the home the National Housing Trust will now give loans to modify, upgrade and install energy efficient equipment and design features. Where two parties are involved with a loan, the amount can be doubled. This loan is actually sufficient to install a solar photovoltaic system, that will provide most of the energy required for the average Jamaican three bedroom household.

In the commercial and the educational sector significant savings can be made in electricity use by motion sensors in rooms. In fact, all commercial infrastructure should be the subject of energy audits and suitable energy management systems put in place. One would like to see a situation where when applying for a loan at a bank, the bank would require an energy audit report to show that the facility is as energy efficient as it could be.

Further, our architects should be trained and stimulated by demand to use passive energy efficient design in order to reduce the demand for energy in a building. This includes, amongst others, orientation so that the sun does not hit windows directly, shading of windows and insulation in walls as well as energy efficient roofing materials.

In the hotel and hospitality sector the use of solar hot water heaters should be pervasive. In addition, one recommends the implementation of an electronic card which opens the door to a guest room being also used to turn on and off the air conditioning and lighting in the room on the entrance and exit of the guest.

The manufacturing sector presents some problems. The cost of electricity to this sector in Jamaica is now nearly 30 cents per kWh. Compare this to the 4.5 US cents per kWh charged for electricity in Trinidad and Tobago. In order to be competitive, Jamaican manufacturers have to be more sensitive to the use of electricity in the most efficient manner. Japan is a model that is worth visiting in that it has relatively high electricity cost but conducts the most efficient use of energy in its manufacturing sector.

Transportation requires significant attention. It is important to discourage the use of high powered engined vehicles which are quite unnecessary given the Jamaican road system and travel destinations. Thus, today we should see an import tariff structure that favours engines less than 200cc as well as high bred vehicles, diesel engined vehicles and ultimately flexi vehicles that can run on essentially 100 per cent ethanol. In the public transportation sector, consideration should be given to modifying the Jamaican taxi vehicles in order for them to use LPG as the fuel. Some 90 per cent of the taxis in Australia run on LPG and 65 per cent of those in Canada do so. Vehicles operating on LPG run cleaner and their engines require less maintenance. However, there is an initial conversion cost of approximately US$3,000.

Supporting these measures in the transport sector will be the rationalization and improvement of urban transportation, the addition of a railway system for the movement of heavy goods and people as well as the consideration of flexi time to avoid the gridlock that obtains on urban roads such as those of the Kingston Metropolitan area during morning and afternoon peak traffic. High fuel prices will dictate a change in our driving habits and lifestyle in respect to travel arrangements. We will now need to become accustomed to reducing driving speeds in order to save fuel, switching lanes less often, driving with windows closed on the highway to avoid wind drag and eliminating quick stop and starts in city traffic. Furthermore we will need to plan our travel needs more carefully in terms of routes to be taken as well as unnecessary trips for domestic errands and other purposes. Car pooling should be an option to be accommodated where possible.

Government has to lead by example in respect to energy efficiency. For this reason, solar water heaters should be used in government buildings such as hospitals wherever and whenever the need for hot water arises. It is important to note that the National Water Commission accounts for approximately 47 per cent of the total energy used by the public sector. Steps have to be taken to make water pumping more efficient because this is an additional cost to the already burdened consumer utility bills.

Notes

Chapter 1: *Energy in the Service of Society*

1. China and Russia serve as excellent examples of the strong international inter-dependence on energy supply. Since the beginning of this century, China has either purchased or developed oil reserves in many parts of the world including Africa, Canada and the United States. Russia has deemphasized if not entirely eliminated foreign investment in its oil and gas sector. However, this has not given any commercial advantage to Russia except that it can now use oil and gas as a political tool. There appears to be no direct economic benefit from this decision because foreign companies would pay just as much tax as the Russian companies.

2. The relationship between having oil resources and "real" wealth is not always clear, it is how the money is spent and if and how the wealth improves the quality of life of the poorest in that society. According to Gilder (1993:57), the flows of oil money do not become an enduring asset of the nation until they can be converted into a stock of remunerative capital – industries, ports, roads, schools, and working skills – that offer a future flow of support when the oil runs out.

Chapter 2: *The Power of Oil*

1. Nevertheless, until the middle of the nineteenth century almost all the lubricated oil derived from animal and vegetable sources and early machines were lubricated with castor and whale oil.

2. The kerosene lamp (widely known in Britain as a paraffin lamp) is any type of lighting device that uses kerosene (paraffin, as distinct from paraffin wax) as a fuel. There are two main types of kerosene lamp which work in different ways, the "wick lamp" and the "pressure lamp".

3. Many independent producers during that time referred to Standard Oil as the 'Old House'.

4. By 1879, Thomas Edison had already developed the mimeograph, the phonograph, storage batteries and motion pictures.
5. This business was dependent on Russian oil, and Marcus Samuel saw an opportunity to diversify the supply, and in 1901 signed a contract to purchase oil at 25 cents per barrel for 25 years.
6. The Royal Dutch Petroleum Company was formed in the Hague in 1890.
7. Mexico became a major producer of oil after the discovery by the American geologist Everette DeGolyner in 1910 of a significant oil source in an area which would be called the Golden Lane, near Tampico. The state oil company, PEMEX, was born after a long battle between government and the oil companies over the stability of agreements and the question of sovereignty and ownership. On March 18, 1938, the expropriation of foreign companies had taken place and was the centrepiece of the Mexican Revolution.
8. These operations were shut down in 1941, during the Second World War.
9. Today, that OPEC is headquartered in Vienna occured purely by accident. It had first established itself in Geneva, but the Swiss doubted its serious intent and did not grant the diplomatic status appropriate to an international organization. The Austrians, eager to obtain the international prestige, were more accommodating, and so, in 1965, OPEC moved to Vienna.

Chapter 3: *Searching for New Petroleum Resources in a Frontier Province*

1. These companies would have provided new seismic and other geophysical data in their offshore acreage.

Chapter 5: *Natural Gas*

1. Nigeria and some other countries have placed a ban on flaring after a certain period in order to obtain maximal value from natural gas sources.
2. The volumetric energy density of CNG is estimated to be 42 per cent of that of LNG and 25 per cent of that of diesel.
3. Natural gas is lighter than air, and disperses quickly when released.
4. Methanol can also be produced from biomass.
5. If global warming occurs, the temperature will rise and decompose some of these methane hydrates in the earth. Methane is one of the most harmful greenhouse gases and this effect could be potentially threatening.

Chapter 6: *Nuclear Power*

1. X-rays had been recently discovered in 1895.
2. Nuclear power plants are some of the most sophisticated and complex energy systems ever designed. However, any complex system, no matter how well it is designed and engineered, cannot be deemed failure-proof (Leuwen et al., 2008).

Chapter 8: *Wind Energy*

1. In 1995 wind energy contributed approximately 10 billion kilowatt hours of electrical generation worldwide, an increase of over 800 per cent since 1980.

Chapter 9: *Solar Power*

1. Central receiver plants use a field of mirrors to focus the sun's energy on a central receiver mounted on a tower. The technology is still experimental.
2. Oliver Headley at the University of the West Indies, Cave Hill, Barbados, has spearheaded important work on solar drying and distillation in the Caribbean and there is a body of literature on these subjects (see Headley, 1994, 1995).

Chapter 10: *Bioenergy*

1. C3 plants are much more common than C4 plants.
2. US researchers have found that switchgrass-derived ethanol produced 540 per cent more energy than was required to manufacture the fuel (BBC). Moreover, unlike corn ethanol, cellulosic biofuel does not require fertilizers, pesticides, energy, and water to grow (State Energy Conservation Office).
3. From Port of Entry, "USA: Cellulosic Ethanol from Sugarcane Bagasse Zooming Ahead to Commercialization". Retrieved from: http://www.portofentry.com/site/root/market/company_news/7020.html
4. Where ethanol is concerned, except for the case of sugar cane, another source of energy is required and which affects prices. The main cost, however, is the raw material, which makes up to 60 to 80 per cent of the total cost of ethanol production (Ahmed, 1994).
5. It is also useful to note that when used in a direct combustion system plant to generate electricity, biomass requires a furnace designed to cope with its higher moisture content and the greater quantity of ash generated.
6. The improved, fuel-efficient biomass stove is also important for millions of people in the developing world. Because of the relatively low cost to implement, stove programmes offer important socioeconomic benefits especially for rural and poor people. The efficient biomass stove can alleviate some of the problems involved in collecting and cooking with fuelwood. The main benefits are that the stoves reduce both the time spent collecting fuel and the indoor air pollution from smoke. Improved charcoal cookstoves have raised energy efficiency in a project in Tanzania by 50 per cent (World Bank, 1994). The two largest improved stove programmes in the world are in India and China, and the lessons culled from these programmes may be instructive to other countries (Barnes and Floor, 1999).
7. Biogas has other uses, but more research is required. Although considerable work has been conducted on the improvement of domestic stoves, the use of biogas for cooking is still inefficient.

8. The government elected castor beans as the preferred feedstock and stimulated the cultivation of castor beans by small producers.

9. Jathropa is not welcome everywhere. Western Australia has banned the plant as invasive and highly toxic to people and animals.

10. Brazil does not have any official legislation on cellulosic ethanol, (cellulosic ethanol is included based on UNICA estimations of market penetration, which amount to around 2.1 billion gallons [12.9per cent] in 2020).

Chapter 12: *Ocean Energy*

1. Nutrients and photosynthesis are primary producers for growth.

2. Reverse osmosis is when fresh water is made out of sea water, also known as desalination.

3. Salt-water evaporation leads to precipitation over land.

Chapter 13: *Geothermal Energy*

1. Direct-use development can heat a single house or a city. Geothermal hot water is piped some 38 kilometres to the city of Reykjavik, Iceland, where it provides heating for the city's 150,000 inhabitants. Near Paris, France, approximately 13,000 apartments are heated by water with a temperature of 60°–80°C from nearby wells between 1,500 and 2,000 metres deep.

2. Geothermal can be used for direct (non-electrical) heating. The technology in this case simply consists of a well, pump and piping.

3. Most of the global geothermal resource is contained in hot dry rock. The pilot project at Fenton Hill, New Mexico, managed by Los Alamos National Laboratory, has been in operation since 1977 investigating the commercial feasibility of hot dry rock geothermal energy. Los Alamos researchers have concluded that the technology is technically feasible and that hot dry rock energy potential at accessible depths could meet all the country's energy needs for more than 5,000 years.

Chapter 15: *Energy Efficiency*

1. The impact of information dissemination is difficult to evaluate.

2. The airline sector has also demonstrated how the elimination of regulated air-fares on air traffic has increased competition, encouraged better utilization, and improved fuel efficiency.

3. The Wigton Wind Farm project is the first major project in Jamaica to benefit from the sale of Carbon Credits under the CDM. Other projects relating to hydropower, wind, solar, biofuels and waste-to-energy projects are under consideration.

Chapter 17: *Future Energy Supply Options*

1. Syngas can be produced from both natural gas and coal, and used as an input to advanced IGCC coal conversion systems or fuel cells. There is no doubt that coal-based IGCC plants will be commercially deployed based on a development programme conducted during the decade of the 1980s (Starr et aI., 1992).

Chapter 18: *Sustainable Energy*

1. Reducing ocean pH and carbonate saturation states.
2. Transport energy is the most rapidly growing source of greenhouse gas emissions and other forms of pollution. Privatization and deregulation are not likely to encourage efficiency, improvement or environmentally acceptable outcomes in the transport sector. Planners are therefore increasingly emphasizing investment in mass transit infrastructure and systems, a policy which would result in regeneration of cities such as Kingston, the capital of Jamaica.
3. Rathmann (2010) discusses the competition between land use for food and liquid biofuels in the context of the Brazillian experience.
4. The technology is presently at the same level of sophistication as the oil and gas industry was about the year 1900.
5. This is a task that is well-suited to the participatory efforts of NGOs.

Epilogue

1. The petroleum product market was liberalized and deregulated in September 1990 and further encouraged by the introduction of the Common External Tariff (CET) in April 1994. This has resulted in an increase in the number of marketing companies retailing petroleum products, greater competition in petroleum pricing at the pump and product diversification, including the addition of brand names to company products.

Glossary

Alternating current (ac): Electric current in which the direction of flow is reversed at frequent intervals; 50 cycles per second is used in Jamaica. This is the current that flows from an electrical outlet.

Alternative Energy: Energy produced from renewable sources, such as sunlight or wind. It has the added benefit of not generating heat trapping "greenhouse" gases.

Ampere: The force of attraction between two parallel current carrying conductors.

Ampere-hour (Amp-hour or Ahr): A measure of electrical charge, equalling the quantity of electricity flowing in one hour past any point of a circuit. Used as a measure of battery capacity.

Anticline: A subsurface geological structure in the form of a sine curve or an elongated dome. The structure is favourable to accumulation of oil and/or gas.

Array: A group of photovoltaic modules wired together to produce a specific amount of power. Array size can range from one to hundreds of modules, depending on how much power will be needed.

Basement rock: Igneous or metamorphic rock lying below the sedimentary formations in the earth's crust. Basement rocks do not contain petroleum deposits.

Biodegradable: Capable of decomposing under natural conditions.

Biofuels: Fuels produced from plant material including trees and grasses and used primarily in engines for transportation.

Carbon dioxide(CO_2): The gas formed in the ordinary combustion of carbon, such as the breathing of animals.

Carbon Sequestration: Carbon sequestration is the storage of carbon dioxide (usually captured from the atmosphere) through biological, chemical or physical processes, for the mitigation of global warming.

Cell (photovoltaic): A semi-conductor device that converts light directly into dc electricity.

CFCs: Chlorofluorocarbons. Used in refrigerants, cleaners and aerosols and in the making of plastic foams. CFCs are greenhouse gases. They also cause ozone depletion in the stratosphere.

Charge controller: A component of a photovoltaic system that controls the flow of current to and from the battery to protect the batteries from over-charge and over-discharge. The charge controller may also indicate the system operational status.

Civil works: Man-made constructions such as buildings, roads, dams, tunnels and cement works of all types.

Compact Fluorescent Bulb: A smaller version of a fluorescent lamp that fits into a standard light bulb socket. Fluorescent bulbs create light in a more energy-efficient way.

Concentrator: A photovoltaic module which includes optical components, such as lenses, to direct and concentrate sunlight onto the small area of a solar cell. Most concentrator arrays must directly face or track the sun.

Condensation: The transformation of a vapour or gas to a liquid by cooling or by increasing pressure on both sides simultaneously.

Cretaceous: The latest period of the Mesozoic Era, between 136 and 65 million years ago (see Tertiary).

Crude oil: Oil as it comes from the well; unrefined petroleum.

Direct current (de): Electric current in which electrons flow in only one direction. This is the current flow produced by a solar system. To be used for typical 120-volt or 220-volt household appliances, it must be converted to alternating current.

Efficiency (of a solar cell or module): The ratio of electric energy produced to the amount of solar energy incident on the cell or module. Crystalline solar modules are about 10 per cent efficient – they convert about 10 per cent of the light energy they receive into electricity.

Energy: The capacity for doing work.

Energy audit: The process of determining the energy consumption of a building or facility.

Energy intensity: The proportion of energy used to gross domestic product (GDP) at constant prices.

Energy security: Energy efficiency, diversity of energy sources, minimizing environmental impact, strengthening energy independence and securing access to safe, affordable energy supplies for all, especially the poor.

Economic Partnership Agreement (EPA): An agreement negotiated between the European Union and other countries such as Cariforum (CARICOM plus Dominican Republic) concerning the trade of goods and services.

Ethanol: An automotive fuel derived from grass, sugar cane or corn. Burning ethanol adds carbon dioxide to the atmosphere, but it is seen as a renewable fuel, like solar power, that does not deplete natural resources.

EU: European Union.

Fault: A fracture in the earth's crust accompanied by the shifting of one side of the fracture with respect to the other.

Flat-plate module or array: A photovoltaic module or array in which the incident solar radiation strikes a flat surface and no concentration of sunlight is involved.

Fluidized beds: Beds of burning fuel together with non-combustible particles kept in suspension by an upward flow of combustion air through the bed. Limestone or coal ash are widely used non-combustible materials.

Gasifiers: Tanks for anaerobic fermentation of biomass residues from sugar cane, pulp and paper, etc., to produce biogas.

Gearbox: A shaft with helical gears that transmits rotor power to generators.

Geothermal energy: Energy produced by the heat in the earth's crust interacting with a water system that produces steam which ultimately drives turbines producing electricity. Geothermal Energy is normally developed in areas around the world which have or have had volcanic activity.

Greenhouse gases: Greenhouse gases are gases in an atmosphere that absorb and emit radiation within the thermal infrared range.

Grid-connected: A photovoltaic system that is connected to a centralized electrical power network.

Hybrid system: A power system consisting of two or more power generating subsystems (e.g., the combination of a wind turbine or diesel generator and a photovoltaic system).

Hydraulic: Operated by the pressure created by forcing water, oil, or another liquid through a comparatively narrow pipe or orifice.

Insolation: The amount of energy in sunlight reaching an area. Usually expressed in watts per square metre (W/m²), but also expressed on a daily basis as watts per square metre per day (W/m²/day).

Inverter: A device that converts direct current (dc) to alternating current (ac) electricity.

Kilowatt (kW): 100 watts.

Kilowatt-hour (kWh): 1,000 watt-hours. A typical residence in Jamaica consumes about 600 kilowatt-hours each month at a price of about US$.14 per kilowatt-hour.

Life cycle cost (LCC) analysis: A form of economic analysis to calculate the total expected costs of ownership over the life span of the system. LCC analysis allows a direct comparison of the costs of alternative energy systems, such as photovoltaics, fossil fuel generators or extending utility power lines.

Load: In an electrical circuit, any device or appliance that uses power (such as a light bulb or water pump).

Maintenance costs: Any costs incurred in the upkeep of a system including replacement and repair of components.

Mariculture: The rearing of fish, as well as other animal and plant life, in the nearshore marine environment. In freshwater conditions the equivalent term is aquaculture.

Module: A number of photovoltaic cells wired together to form a unit, usually in a sealed frame of convenient size for handling and assembling into arrays also called a 'panel'.

Ohm: The unit of electrical resistance equal to the resistance through which a current of 1 ampere will flow when there is a potential difference of one 1 volt across it.

Operating costs: The costs of using a system. For fuel-based systems these costs include all fuel costs over the system's lifetime.

Parabolic reflector: A reflective device, commonly formed in the shape of a paraboloid of revolution. Parabolic reflectors can either collect or distribute energy such as light, sound, or radio waves.

Parallel connection: A wiring configuration used to increase current (amperage). Parallel wiring is positive to positive (+ to +) and negative to negative (– to –). Opposite of a series connection.

Peak watts (Wp): The maximum power (in watts) a solar array will produce on a clear, sunny day while the array is in full sunlight and operating at 25°C. Actual wattage at higher temperatures is usually somewhat lower.

Photovoltaics: The use of lenses or mirrors to concentrate direct solar radiation onto small areas of solar cells, or the use of flat-plate photovoltaic modules using large arrays of solar cells, to convert sunlight into electricity.

Photovoltaic (PV) system: A complete set of interconnected components for converting the sun's radiation into electricity by the photovoltaic process, including array, other components, and the load.

Power: The rate at which energy is consumed or generated. Power is measured in watts or horsepower.

Power conditioner: The electrical equipment used to convert power from a photovoltaic array into a form suitable for use. Loosely, a collective term for inverter, transformer, voltage regulator, and other power controls.

Prime mover: Any source of motion such as engines or electric motors.

Renewable energy: Flows of energy that are regenerative or virtually inexhaustible. Most commonly includes solar (electricity and thermal), biomass, geothermal, wind, tidal, wave, and hydro power sources.

Reservoir: Permeable sedimentary rock formation containing quantities of oil and/or gas enclosed or surrounded by layers of less permeable rock forming a structural or stratigraphic trap.

Royalty: A share in the gross production of a mineral such as oil and gas in a property or country without bearing any of the production cost.

Sedimentary rock: Rock formed by material laid down by seas, rivers or lakes. Limestones, sandstone, shale are sedimentary rocks.

Seismograph: A device that records vibrations from the earth. As used in oil and gas exploration a seismograph records shock waves set off by explosions detonated in shot holes and picked up by geophones.

Series connection: A wiring configuration used to increase voltage. Series wiring is positive to negative (+ to −) or negative to positive (− to +). Opposite of a parallel connection.

Silicon: A non-metallic element that, when specially treated, is sensitive to light and capable of transforming light into electricity. Silicon is the basic material of beach sand, and is the raw material used to manufacture most photovoltaic cells.

Sinks: Places where carbon dioxide is absorbed – the oceans, soil and detritus, trees and other vegetation.

Stand-alone photovoltaic system: A solar electric system commonly used in remote locations, not connected to the main electric grid. Most stand-alone systems include some type of energy storage, such as batteries or pumped water.

Quaternary Tertiary	Cenozoic 65M Years
Cretaceous Jurassic Triassic	Mesozoic 225M Years
Permian Carboniferous Devonian Silurian Ordovician Cambrian	Paleozoic 600M Years
Precambrian	

Chart showing the periods and eras of geologic time.

Tertiary: The older major period of the Cenozoic era, extending from the end of the Cretaceous to the beginning of the Quaternary, from 65 million to 2 million years ago.

Vapour pressure: The pressure exerted by a vapour held in equilibrium with its liquid state. Stated inversely, it is the pressure required to prevent a liquid from changing to a vapour.

Voltage: A measure of the force or push given the electrons in an electrical circuit; a measure of electric potential. One volt produces one ampere of current when acting against a resistance of one ohm.

Von Post scale is the most widely used system for determining the degree of decomposition of peat. The degree of decomposition is called the H-value and is expressed by a number on a scale that ranges from 1 to 10, with H-1 being totally undecomposed plant material and H-10 completely decomposed peat. Each point represents about 10 per cent decomposition. The degree of decomposition is determined by squeezing freshly harvested peat in the hand and examining the compressed peat and water. This method is very useful for assessing sphagnum peat but not as suitable for sedge and woody peat.

Watt (W): A measure of electric power in a unit of time, equal to the rate of flow (amps) multiplied by the voltage of that flow (volts). One amp of current flowing at a potential of one volt produces one watt of power.

Watt-hour (Wh): A measure of electrical energy equal to the amount of electrical power multiplied by the length of time (hours) the power is applied.

Wildcat well: A well drilled in an unproved area; far from a producing well.

Energy Conversion Factors and Equivalents

Basic Energy Units

1 joule (J) = 0.2388 cal

1 calorie (cal) = 4.1868 J

(1 British thermal unit [Btu] = 1.055 kJ = 0.252 kcal)

WEC Standard Energy Units

1 tonne of oil equivalent (toe) = 42 GJ (net calorific value) = 10,034 Mcal

1 tonne of coal equivalent (tce) = 29.3 GJ (net calorific value) = 7,000 Mcal

Note: the tonne of oil equivalent currently employed by the International Energy Agency and the United Nations Statistics Division is defined as 107 kilocalories, net calorific value (equivalent to 41.868 GJ).

Volumetric Equivalents

1 barrel = 42

US gallons = approx. 159 litres

1 cubic metre = 35.315 cubic feet = 6.2898 barrels

Electricity

1 kWh of electricity output = 3.6 MJ = approx. 860 kcal

Representative Average Conversion Factors

1 tonne of crude oil = approx. 7.3 barrels

1 tonne of natural gas liquids = 45 GJ (net calorific value)

1,000 standard cubic metres of natural gas = 36 GJ (net calorific value)

1 tonne of uranium (light-water reactors, open cycle) = 10,000–16,000 toe

1 tonne of peat = 0.2275 toe

1 tonne of fuel wood = 0.3215 toe

1 kWh (primary energy equivalent) = 9.36 MJ = approx. 2,236 Mcal

Index

www.ingramcontent.com/pod-product-compliance
Lightning Source LLC
Chambersburg PA
CBHW080519220326
41599CB00032B/6134